Smectic Liquid Crystals—Textures and Structures

To our respective wives—Marjorie and Ann—who have too often, with our families, been neglected in the cause of liquid crystals, and to whom therefore this book owes a great deal.

SMECTIC LIQUID CRYSTALS
Textures and Structures

G.W. GRAY, PhD, FRSC, FRS

Professor of Organic Chemistry
University of Hull, UK

and

J.W.G. GOODBY, PhD, MRSC

AT & T Bell Laboratories, New Jersey, USA

Leonard Hill

Glasgow and London
Distributed in the USA and Canada by
Heyden & Son, Inc.
Philadelphia

CHEMISTRY

Published by Leonard Hill
A member of the Blackie Group
Bishopbriggs
Glasgow G64 2NZ

Furnival House
14–18 High Holborn
London WC1V 6BX

Distributed in the USA and Canada by
Heyden & Son, Inc.
247 South 41st Street,
Philadelphia, PA 19104

© 1984 G.W. Gray and J.W. Goodby
First Published 1984

All rights reserved.
No part of this publication may be reproduced,
stored in a retrieval system, or transmitted,
in any form or by any means,
electronic, mechanical, recording or otherwise,
without the prior permission of the Publishers.

British Library Cataloguing in Publication Data
Gray, G.W.
  Smectic liquid crystals—textures and structures.
  1. Liquid crystals
  I. Title   II. Goodby, J.W.
  548'.9   QD923

ISBN 0–249–44168–3

For the USA and Canada,
International Standard Book Number is
0–86344–025–8

Printed in Great Britain by McCorquodale (Scotland) Ltd.

Preface

This book is intended as a practical and useful experimental guide to the textures and classification of smectic liquid crystals of different polymorphic types. The period of writing and construction of the book has coincided with a particularly active era during which knowledge of smectic systems has evolved at a rapid rate. As this has happened, it has been necessary to revise the original text, introduce new sections, and extend the examples of textures given by the photomicrographs. This activity has consumed time, but we hope that the result is a superior text with better coverage of the subject. Even so, it became clear that some areas of development still needed discussion, but that this would be difficult to integrate into the existing text. An appendix up-dating these topics (Chapter 10) has been included as a practical way of overcoming this difficulty.

As elaborated in the Introduction, this book is intended not simply for those already knowledgeable in the field, but also for the research worker who is beginning work in the complex area of smectics and wishes to use optical microscopy for the purposes of phase identification. This is, of course, a subject in which experience counts for a great deal. The authors hope to pass on to relative newcomers to the field some of their own experience, and to provide them with reference photographs of textures that are *typical* of those that may be encountered in novel materials. In electing to portray typical textures, this book does differ somewhat from another (Demus and Richter, 1978) dealing with liquid crystal textures, in which considerable emphasis is placed on the effects of structural disclinations upon texture and the origins of these in local deformations of or discontinuities in the arrangement of the molecules. This aspect is important, but we judged that the average research worker is first concerned with what is typical. The detailed effects of deformations on texture may be considered with benefit later, and in this respect, the two books are complementary.

This book was conceived several years ago when the two authors worked in collaboration at the University of Hull. It has progressed over the years, during which one of us (JWG) moved to the United States.

In compiling this book, we have been helped in many ways, and wish to express particular gratitude to Mrs R. Knight for typing the text, Mr A.T. Rendell for his valued work in producing the colour-true photographic prints of textures, Professor A.J. Leadbetter of the Rutherford Appleton Laboratories for many helpful discussions, Professor H. Sackmann, Professor D. Demus, and their colleagues of the University of Halle, East Germany for their co-operation in reaching, with us, an agreed system of nomenclature for some of the smectic modifications, and the Department of Chemistry of the University of Hull, England, for making available facilities without which this book could not have been completed.

G.W.G.
J.W.G.

Reference

Demus, D. and Richter, L. (1978). *The Textures of Liquid Crystals.* V.E.B. Deutscher Verlag für Grundstoffindustrie, Leipzig.

Contents

Plate legends in numerical order ix

Plate numbers for specific phase types xvii

Introduction xix

1 The smectic A phase

Introduction	1
Structure of the smectic A phase	1
The bilayer smectic A phase	6
Textures of the smectic A phase	8
Identification and classification of the smectic A phase	19

2 The smectic B phase

Introduction	23
Structure of the smectic B phase	25
Textures of the smectic B phase	35
Identification and classification of the smectic B phase	42

3 The smectic C phase

Introduction	45
Structure of the smectic C phase	46
Theories of the smectic C phase	48
Textures of the smectic C phase	57
The chiral smectic C phase	60
Identification and classification of the smectic C phase	64

4 The smectic D phase

Introduction	68
Structure of the smectic D phase	69
Textures of the smectic D phase	78
Identification and classification of the smectic D phase	80

5 The smectic E phase

Introduction	82
Structure of the smectic E phase	83
Textures of the smectic E phase	88
Identification and classification of the smectic E phase	91

6 The smectic F phase

Introduction	94
Structure of the smectic F phase	95
The chiral smectic F phase	99
Textures of the smectic F phase	99
Identification and classification of the smectic F phase	102

7 The smectic G phase

Introduction	105
Structure of the smectic G phase	107
Structure of the 'chiral' smectic G phase	110
Textures of the smectic G phase	112
Identification and classification of the smectic G phase	116

8 The smectic H phase

Introduction	120
Structure of the smectic H phase	120
Textures of the smectic H phase	123
Identification and classification of the smectic H phase	125

9 The smectic I phase

Introduction	127
Structure of the smectic I phase	129
Textures of the smectic I phase	131
Identification and classification of the smectic I phase	132

10 Some new developments in phase classification and structure

Introduction	134
Hexatic and crystal smectic phases	134
Classification of hexatic and crystal B phases	136
Textures of the crystal and hexatic B phase	138
Tilted hexatic and crystal phases	139
Identification and classification of crystal and hexatic B phases	141
Antiphase behaviour	143
Smectic B phases	148
Ferroelectric phases	149
The smectic F to isotropic liquid transition	151
Structural features of smectic phases	153

General index   159

Plates 1–124   165

List of plates

Plate 1	The focal-conic fan texture of the smectic A phase of n-decyl 4-(4′-phenyl-benzylideneamino)cinnamate.
Plate 2	The separation of the smectic A phase in the form of bâtonnets from the isotropic liquid of diethyl 4,4′-azoxydibenzoate.
Plate 3	The polygonal texture of the smectic A phase of n-butyl 4-(4′-phenylbenzyl-ideneamino)cinnamate.
Plate 4	The natural texture of the smectic B phase obtained on cooling the isotropic liquid of 4-n-hexyl-4′-n-hexyloxybiphenyl; the black areas are homeotropic S_B.
Plate 5	The transition to the mosaic texture of the smectic B phase on cooling the nematic phase of 4-n-pentyloxybenzylidene-4′-aminobiphenyl.
Plate 6	The mosaic texture of the smectic B phase of 4-n-pentyloxybenzyl-idene-4′-aminobiphenyl (oblong platelets).
Plate 7	The mosaic texture of the smectic B phase of n-propyl 4-(4′-propylmercapto-benzylideneamino)cinnamate (more rounded platelets).
Plate 8	The paramorphotic focal-conic fan texture of the ordered smectic B phase of n-decyl 4-(4′phenylbenzylideneamino)cinnamate formed on cooling the smectic A phase shown in Plate 1.
Plate 9	The paramorphotic focal-conic fan texture of the hexatic, fluid smectic B phase of n-hexyl 4′-n-pentyloxybiphenyl-4-carboxylate.
Plate 10	The paramorphotic truncated focal-conic fan texture of the smectic B phase formed on cooling the smectic A phase of N-(4′-n-heptyloxy-benzylidene)-4-ethylaniline (7O.2).
Plate 11	The transition from the focal-conic fan texture of the smectic A phase to the paramorphotic focal-conic fan texture of the smectic B phase for n-decyl 4-(4′phenylbenzylideneamino)cinnamate. The backs of the fans are crossed with transition bars.
Plate 12	More pronounced transition bars occurring at the smectic A to smectic B phase change in a mixture of terephthalylidene-bis-4-n-butylaniline (50% by wt) and n-hexyl 4′-n-dodecyloxybiphenyl-4-carboxylate (50% by wt).
Plate 13	The paramorphotic focal-conic fan texture (also showing some areas with a moss-like appearance) of the smectic B phase formed after the isotropic liquid to S_{AB} transition for 3-methylbenzyl 4-(4′-phenylbenzylideneamino)cinna-mate.
Plate 14	The *separation* of the smectic B phase from the isotropic liquid *via* an infinitely short smectic A phase for 3-methylbenzyl 4-(4′phenylbenzylideneamino) cinnamate—compare with Plate 13 which shows the texture on completion of the transition.

Plate 15	The transition from the isotropic liquid to the S_A to the S_B phase for methyl 4′-n-octyloxybiphenyl-4-carboxylate. The S_A phase has a small, but more finite temperature range than that for the ester in Plate 13 and 14.
Plate 16	The natural texture of the smectic B phase separating from the nematic phase of *trans, trans*-4-n-propylbicyclohexyl-4′-carbonitrile on slow cooling.
Plate 17	The large platelets of the mosaic texture of the smectic B phase formed on cooling the nematic phase of *trans, trans*-4-n-propylbicyclohexyl-4′-carbonitrile.
Plate 18	The natural *schlieren* texture of the smectic C phase formed on cooling the nematic phase of $(\pm)$-4-(2′-methylbutyl)phenyl 4′-(4″-methylhexyl)-biphenyl-4-carboxylate.
Plate 19	The *schlieren* texture of the smectic C phase formed on cooling the homeotropic texture of the smectic A phase of 4-n-hexyloxyphenyl 4′-n-octyloxybiphenyl-4-carboxylate. The texture exhibits only centres with four *schlieren*.
Plate 20	The sanded texture of the smectic C phase of 3′-nitro-4′-n-hexadecyloxybiphenyl-4-carboxylic acid.
Plate 21	The *schlieren* texture of the smectic C phase of 4-n-hexyloxyphenyl 4′-n-octyloxybiphenyl-4-carboxylate. The small domains exhibit *faint* lines caused by layer stress as the tilt angle changes.
Plate 22	The paramorphotic broken focal-conic fan texture of the smectic C phase formed on cooling the smectic A phase of terephthalylidene-bis-4-n-butylaniline.
Plate 23	Contact preparation of chiral and non-chiral smectic C phases. A racemic mixture of two optically active isomers of 4-(2′-methylbutyl)phenyl 4′-n-octyloxybiphenyl-4-carboxylate (top of plate) is allowed to form a contact preparation with S-(+)-4-(2′-methylbutyl)phenyl 4′-n-octyloxybiphenyl-4-carboxylate (lower part of plate).
Plate 24	The petal texture of the chiral smectic C phase of S-(+)-4-n-hexyloxyphenyl 4′-(4″-methylhexyl)biphenyl-4-carboxylate.
Plate 25	The texture of the chiral smectic C phase when the plane layers dip at an angle to the glass surface for S-(+)-terephthalylidene-bis-4-(4′-methylhexyloxy)aniline. The equally spaced lines are related to the pitch of the phase.
Plate 26	Optically isotropic areas of the smectic D phase forming in the sanded texture of the smectic C phase on heating 3′-nitro-4′-n-hexadecyloxybiphenyl-4-carboxylic acid.
Plate 27	The fern growth pattern of the smectic D phase forming in the *schlieren* C texture of 3′-nitro-4′-n-hexadecyloxybiphenyl-4-carboxylic acid.
Plate 28	The fern-like growth pattern of the smectic D phase of 3′-nitro-4′-n-hexadecyloxybiphenyl-4-carboxylic acid. The fern pattern demonstrates the growth of the cubic or octahedral domains as they mesh together to form a single domain.
Plate 29	The texture of the S_4 or discotic phase formed on cooling the smectic A phase of 3′-nitro-4′-n-hexadecyloxybiphenyl-4-carboxylic acid.
Plate 30	The natural texture of the smectic E phase separating from the isotropic liquid on cooling 4-ethoxy-4′-acetylbiphenyl.
Plate 31	The natural mosaic texture of the smectic E phase obtained on cooling the isotropic liquid of 4-ethoxy-4′-acetylbiphenyl.
Plate 32	The paramorphotic arced focal-conic fan texture of the smectic E phase obtained on cooling the focal-conic textures of the smectic A and smectic B phases of methyl 4′-n-octyloxybiphenyl-4-carboxylate (see also Sequence 7).
Plate 33	The platelet texture of the smectic E phase obtained on cooling the homeotropic texture of the smectic A phase of di-n-propyl *p*-terphenyl-4,4″-carboxylate.

LIST OF PLATES

Plate 34 The paramorphotic arced focal-conic fan texture of the smectic E phase (again showing moss-like areas) obtained on cooling the focal-conic-mosaic texture obtained after the isotropic liquid to S_{AB} transition for 3-methylbenzyl 4-(4'-phenylbenzylideneamino)cinnamate.

Plate 35 The paramorphotic broken focal-conic fan texture of the smectic F phase formed on cooling the focal-conic texture of the smectic C phase of terephthalylidene-bis-4-n-pentylaniline (TBPA).

Plate 36 The paramorphotic *schlieren*-mosaic texture of the smectic F phase formed on cooling the *schlieren* texture of the smectic C phase of terephthalylidene-bis-4-n-pentylaniline (TBPA).

Plate 37 The paramorphotic broken focal-conic fan texture of the smectic F phase formed on cooling the smectic A phase of N-(4-n-nonyloxybenzylidene)-4'-n-butylaniline (9O.4).

Plate 38 The natural mosaic texture of the smectic F phase formed on cooling the homeotropic texture of the smectic A phase of N-(4-n-nonyloxybenzylidene)-4'-n-butylaniline (9O.4).

Plate 39 The very fine mosaic texture of the smectic F phase of N-(4-n-pentyloxybenzylidene)-4'-n-hexylaniline (5O.6) formed on cooling the homeotropic texture of the preceding smectic B phase.

Plate 40 The broken focal-conic texture of the chiral smectic F phase of (+)-4-(2"-chlorobutanoyloxy)-4'-n-octyloxybiphenyl.

Plate 41 The *schlieren*-mosaic texture of the chiral smectic F phase of (+)-4-(2"-chlorobutanoyloxy)-4'-n-octyloxybiphenyl.

Plate 42 The smectic G phase of (±)-4-(2"-chlorobutanoyloxy)-4'-n-pentyloxybiphenyl separating from the isotropic liquid on cooling.

Plate 43 The natural mosaic texture of the smectic G phase of (±)-4-(2"-chlorobutanoyloxy)-4'-n-pentyloxybiphenyl formed on cooling the isotropic liquid.

Plate 44 The mosaic texture of the smectic G phase separating from the nematic phase of N-(4-n-butyloxybenzylidene)-4'-ethylaniline (4O.2) on rapid cooling.

Plate 45 The paramorphotic broken focal-conic fan texture of the smectic G phase formed on cooling the fan texture of the smectic C phase of terephthalylidene-bis-4-n-butylaniline (TBBA).

Plate 46 The mosaic texture of the smectic G phase formed on cooling the *schlieren* texture of the smectic C phase of terephthalylidene-bis-4-n-butylaniline (TBBA).

Plate 47 The texture of the 'chiral' smectic G phase formed on cooling the *schlieren* texture of the chiral smectic C phase of (+)-4-(2'-methylbutyl)phenyl 4'-n-octyloxybiphenyl-4-carboxylate. This texture is of a mosaic kind, but retains *schlieren* characteristics from the smectic C phase.

Plate 48 Immiscibility of the smectic G mosaic texture and the smectic B homeotropic texture in a thoroughly mixed, two-component mixture of terephthalylidene-bis-4-n-butylaniline (80% by wt) (S_G) and 4-n-pentyloxyphenyl 4'-n-octyloxybiphenyl-4-carboxylate (20% by wt) (S_B). The two phases segregate on cooling the continuous and homogeneous *schlieren* texture of the preceding smectic C phase.

Plate 49 The paramorphotic broken focal-conic fan and mosaic textures of the smectic H'(K) phase formed on cooling the fan and mosaic textures of the smectic G phase of (±)-4-(2'-methylbutyl)phenyl 4'-n-decyloxybiphenyl-4-carboxylate.

Plate 50 The paramorphotic broken focal-conic fan texture of the smectic H'(K) phase of (+)-4-(2'-methylbutyl)phenyl 4'-n-octyloxybiphenyl-4-carboxylate.

Plate 51	The mosaic texture of the smectic H phase with zig-zag lines formed from the mosaic texture of the smectic G phase of terephthalylidene-bis-4-n-butylaniline (TBBA).
Plate 52	The grained paramorphotic mosaic texture of the smectic H phase formed on cooling the mosaic texture of the smectic G phase of terephthalylidene-bis-4-n-pentylaniline (TBPA).
Plate 53	The natural texture of the smectic I phase formed on cooling the isotropic liquid of 4,4'-bis-(n-octadecylamino)biphenyl.
Plate 54	The *schlieren* texture of the smectic I phase formed on cooling the *schlieren* texture of the smectic C phase of ($\pm$)-4-(2'-methylbutyl)phenyl 4'-n-nonyloxybiphenyl-4-carboxylate.
Plate 55	The bubble or plane texture of the smectic I phase of (+)-4-(2'-methylbutyl)phenyl 4'-n-octyloxybiphenyl-4-carboxylate formed on cooling the plane texture of the preceding chiral smectic C phase.
Plate 56	The focal-conic texture of the chiral smectic I phase of (+)-4-(2'-methylbutyl)phenyl 4'-n-octyloxybiphenyl-4-carboxylate.
Plate 57	A contact preparation between the optically active smectic I phase and its achiral counterpart. The material is racemic and S-(+)-4-(2'-methylbutyl)phenyl 4'-n-octyloxybiphenyl-4-carboxylate.
Plate 58	Unidentified S_{VII} phase of terephthalylidene-bis-4-n-butylaniline (TBBA) formed on cooling the mosaic texture of the preceding smectic H phase.
Plate 59	The unidentified S_2 phase of *trans, trans*-4-n-propylbicyclohexyl-4'-carbonitrile formed on cooling the mosaic texture of the preceding S_B phase.
Plate 60	The second unidentified S_3 phase of *trans, trans*-4-n-propylbicyclohexyl-4'-carbonitrile formed on cooling the mosaic texture of the preceding S_2 phase.
Plate 61	*Sequence 1.* The *schlieren* texture of the smectic C phase of racemic 4-(2'-methylbutyl)phenyl 4'-n-octyloxybiphenyl-4-carboxylate.
Plate 62	*Sequence 1.* The *schlieren* texture of the smectic C phase on further cooling—note the colour changes as the tilt angle increases with decreasing temperature—for racemic 4-(2'-methylbutyl)phenyl 4'-n-octyloxybiphenyl-4-carboxylate.
Plate 63	*Sequence 1.* The S_C to S_I phase transition on cooling for racemic 4-(2'-methylbutyl)phenyl 4'-n-octyloxybiphenyl-4-carboxylate (same area as Plates 61 and 62).
Plate 64	*Sequence 1.* The *schlieren* texture of the smectic I phase of racemic 4-(2'-methylbutyl)phenyl 4'-n-octyloxybiphenyl-4-carboxylate (same area as previous Plates).
Plate 65	*Sequence 1.* The mosaic texture of the smectic G'(J) phase of racemic 4-(2'-methylbutyl)phenyl 4'-n-octyloxybiphenyl-4-carboxylate (same area as previous Plates).
Plate 66	*Sequence 1.* The transition from the mosaic texture of the smectic G'(J) phase to the smectic H'(K) phase of racemic 4-(2'-methylbutyl)phenyl 4'-n-octyloxybiphenyl-4-carboxylate (same area as Plate 65).
Plate 67	*Sequence 1.* The mosaic texture of the smectic H'(K) phase of racemic 4-(2'-methylbutyl)phenyl 4'-n-octyloxybiphenyl-4-carboxylate (same area as Plate 66).
Plate 68	*Sequence 2.* The focal-conic fan texture of the smectic A phase of S-(+)-4-(2'-methylbutyl)phenyl 4'-n-octyloxybiphenyl-4-carboxylate.
Plate 69	*Sequence 2.* The banded focal-conic fan texture of the chiral smectic C phase of S-(+)-4-(2'-methylbutyl)phenyl 4'-n-octyloxybiphenyl-4-carboxylate.

LIST OF PLATES xiii

Plate 70 *Sequence 2.* The broken focal-conic fan texture of the chiral smectic I phase of S-(+)-4-(2'-methylbutyl)phenyl 4'-n-octyloxybiphenyl-4-carboxylate.

Plate 71 *Sequence 2.* The broken, banded focal-conic fan texture of the 'chiral' smectic G'(J) phase of S-(+)-4-(2'-methylbutyl)phenyl 4'-n-octyloxybiphenyl-4-carboxylate (same area as Plates 68, 69, and 70).

Plate 72 *Sequence 3.* The focal-conic fan and homeotropic textures of the smectic A phase of N-(4-n-heptyloxybenzylidene)-4'-n-pentylaniline (7O.5).

Plate 73 *Sequence 3.* The broken focal-conic fan and *schlieren* textures of the smectic C phase of N-(4-n-heptyloxybenzylidene)-4'-n-pentylaniline (7O.5).

Plate 74 *Sequence 3.* The focal-conic fan and homeotropic textures of the crystal smectic B phase of N-(4-n-heptyloxybenzylidene)-4'-n-pentylaniline (7O.5).

Plate 75 *Sequence 3.* The arced or banded focal-conic fan and mosaic textures of the smectic G phase of N-(4-n-heptyloxybenzylidene)-4'-n-pentylaniline (7O.5).

Plate 76 *Sequence 4.* The focal-conic fan texture of the crystal smectic B phase of N-(4-n-pentyloxybenzylidene)-4'-n-hexylaniline (5O.6).

Plate 77 *Sequence 4.* The broken focal-conic fan texture of the smectic F phase of N-(4-n-pentyloxybenzylidene)-4'-n-hexylaniline (5O.6).

Plate 78 *Sequence 4.* The broken focal-conic fan texture of the smectic G phase of N-(4-n-pentyloxybenzylidene)-4'-n-hexylaniline (5O.6).

Plate 79 *Sequence 5.* The *schlieren* and focal-conic fan textures of the smectic I phase formed on cooling the C phase of N, N'-bis-(4'-n-heptyloxybenzylidene)-1,4-phenylenediamine. Note the focal-conic fan texture is relatively unbroken because the C phase was formed directly from the nematic phase. This textural property is carried through all of the phases (Plates 80, 81, and 82).

Plate 80 *Sequence 5.* The mosaic and focal-conic fan textures of the smectic G'(J) phase formed on cooling the smectic I phase of N, N'-bis-(4'-n-heptyloxybenzylidene)-1,4-phenylenediamine.

Plate 81 *Sequence 5.* The transition textures at the smectic G'(J) to smectic H'(K) phase change of N, N'-bis-(4'-n-heptyloxybenzylidene)-1,4-phenylenediamine.

Plate 82 *Sequence 5.* The mosaic and focal-conic fan textures of the smectic H'(K) phase of N, N'-bis-(4'-n-heptyloxybenzylidene)-1,4-phenylenediamine.

Plate 83 *Sequence 6.* The focal-conic fan and homeotropic textures of the smectic A phase of terephthalylidene-bis-4-n-decylaniline (TBDA).

Plate 84 *Sequence 6.* The *schlieren* and broken focal-conic fan textures of the smectic C phase of terephthalylidene-bis-4-n-decylaniline (TBDA).

Plate 85 *Sequence 6.* The *schlieren* and focal-conic fan textures of the smectic I phase of terephthalylidene-bis-4-n-decylaniline (TBNA).

Plate 86 *Sequence 6.* The mosaic and focal-conic fan textures of the smectic F phase of terephthalylidene-bis-4-n-decylaniline (TBDA).

Plate 87 *Sequence 7.* The smectic A phase of n-hexyl 4'-n-pentyloxybiphenyl-4-carboxylate (65OBC), formed on cooling the isotropic liquid.

Plate 88 *Sequence 7.* The hexatic B phase of n-hexyl 4'-n-pentyloxybiphenyl-4-carboxylate (65OBC) formed on cooling the focal-conic texture of the preceding A phase.

Plate 89 *Sequence 7.* The smectic E phase of n-hexyl 4'-n-pentyloxybiphenyl-4-carboxylate (65OBC) typically showing a banded focal-conic texture.

Plate 90 *Sequence 7.* The smectic A phase of n-hexyl 4'-n-pentyloxybiphenyl-4-carboxylate (65OBC) formed on heating the hexatic B phase. The backs of the focal-conic domains become crossed with parabolic focal-conic defects (wishbones)—see also Sequence 8.

Plate 91   *Sequence 8.* The focal-conic fan texture of the smectic A phase of n-hexyl 4′-n-pentylbiphenyl-4-thiolcarboxylate (65SBC).

Plate 92   *Sequence 8.* The transition from the A phase to the crystal B phase of n-hexyl 4′-n-pentylbiphenyl-4-thiolcarboxylate (65SBC). The fan backs become crossed with transition bars.

Plate 93   *Sequence 8.* The focal-conic fan texture of the crystal B phase of n-hexyl 4′-n-pentylbiphenyl-4-thiolcarboxylate (65SBC).

Plate 94   *Sequence 8.* The arced focal-conic fan texture of the smectic E phase formed on cooling the crystal B phase of n-hexyl 4′-n-pentylbiphenyl-4-thiolcarboxylate (65SBC).

Plate 95   *Sequence 8.* The smectic A phase of n-hexyl 4′-n-pentylbiphenyl-4-thiolcarboxylate (65SBC) formed on heating the crystal B phase. The backs of the focal-conic domains become patterned with parabolic focal-conic defects.

Plate 96   The crystal B phase of n-propyl 4′-n-pentylbiphenyl-4-thiolcarboxylate (35SBC) formed on cooling the focal-conic fan texture of the preceding A phase. Note the fans now have a stepped appearance; compare with Plate 88 for the hexatic B phase.

Plate 97   *Sequence 9.* The homeotropic and focal-conic fan textures of the bilayer smectic A phase of 4′-n-nonyloxy-4-biphenylyl 4-cyanobenzoate.

Plate 98   *Sequence 9.* The transition from the bilayer smectic A phase to the tilted antiphase ($S_{\tilde{C}}$) of 4′-n-nonyloxy-4-biphenylyl 4-cyanobenzoate.

Plate 99   *Sequence 9.* The textures of the tilted antiphase ($S_{\tilde{C}}$) of 4′-n-nonyloxy-4-biphenylyl 4-cyanobenzoate.

Plate 100   The transition from the homeotropic texture of the bilayer smectic A phase to the domain texture of the tilted antiphase ($S_{\tilde{C}}$) of 4′-n-nonyloxy-4-biphenylyl 4-cyanobenzoate.

Plate 101   The domain texture of the tilted antiphase ($S_{\tilde{C}}$) of 4′-n-decyloxy-4-biphenylyl 4-cyanobenzoate.

Plate 102   *Sequence 10.* The smectic A phase of 4-n-heptylphenyl 4-(4′-nitrobenzoyloxy)benzoate.

Plate 103   *Sequence 10.* The transition from the smectic A phase to the smectic A antiphase ($S_{\tilde{A}}$) of 4-n-heptylphenyl 4-(4′-nitrobenzoyloxy)benzoate.

Plate 104   *Sequence 10.* The smectic A antiphase ($S_{\tilde{A}}$) of 4-n-heptylphenyl 4-(4′-nitrobenzoyloxy)benzoate.

Plate 105   The smectic F phase separating from the isotropic liquid of 4-dodecanoyloxy-4′-n-octyloxybiphenyl.

Plate 106   The *schlieren* texture of the nematic phase of 4-(2′-methylbutyl)phenyl 4′-(4″-methylhexyl)biphenyl-4-carboxylate.

Plate 107   The threaded texture of the nematic phase of 4-(*trans*-4′-n-pentylcyclohexylmethoxy)benzonitrile showing two brushes originating from the centres.

Plate 108   The *schlieren* nematic texture of 4-(*trans*-4′-n-pentylcyclohexylmethoxy) benzonitrile showing four brushes radiating from several centres.

Plate 109   Plane texture of the cholesteric phase (with discontinuities) of chiral di-2-methylbutyl terephthalylidene-bis-4′-aminobenzoate.

Plate 110   The focal-conic texture of the cholesteric phase of chiral 2-methylbutyl 4-(4′-nitrobenzylideneamino)cinnamate.

LIST OF PLATES

Plate 111	Rivulet of nematic in a contact preparation of the plane textures of two cholesteric materials of opposite pitch sense.
Plate 112	Continuity in a contact preparation of the plane textures of two cholesteric materials of the same pitch sense.
Plate 113	The blue fog phase of (+)-4'-n-hexyloxy-4-biphenylyl 4-(2'-methylbutyl)benzoate separating from the isotropic liquid.
Plate 114	The platelet texture of 'blue' phase (II) of (+)-4'-n-hexyloxy-4-biphenylyl 4-(2'-methylbutyl)benzoate formed on cooling the blue fog phase.
Plate 115	The wrinkled platelet texture of 'blue' phase (I) of (+)-4'-n-hexyloxy-4-biphenylyl 4-(2'-methylbutyl)benzoate formed on cooling 'blue' phase (II).
Plate 116	The zig-zag blade texture of the cholesteric 'blue' phase of (+)-4-(2'-methylbutyl)phenyl 4'-(4"-methylhexyl)biphenyl-4-carboxylate.
Plate 117	The discotic phase of benzene-hexa-n-heptanoate separating from the isotropic liquid on cooling.
Plate 118	The discotic phase of benzene-hexa-n-heptanoate reverting to the isotropic liquid on heating.
Plate 119	The pseudo focal-conic texture of the discotic phase of benzene-hexa-n-heptanoate.
Plate 120	The feather texture of the discotic phase of benzene-hexa-n-heptanoate.
Plate 121	The pseudo focal-conic fan texture of the discotic phase of di-isobutylsilanediol.
Plate 122	The spine texture of the discotic phase of di-isobutylsilanediol.
Plate 123	The spine texture of the discotic phase of di-isobutylsilanediol separating from the isotropic liquid.
Plate 124	The texture of the discotic phase of Uroporphyrin (I) octa-n-dodecyl ester separating from the isotropic liquid.

Plate numbers for specific phase types

'Blue' phase	Plates 113–116	Smectic $\tilde{C}$	Plates 98–101
Cholesteric	Plates 109–112	Smectic D	Plates 26–28
Discotic	Plates 29, 117–124	Smectic E	Plates 30–34, 89, 94
Nematic	Plates 106–108	Smectic F	Plates 35–39, 77, 86, 105
Smectic A	Plates 1–3, 11, 12, 68, 72, 83, 87, 90–92, 95, 102	Smectic F*	Plates 40, 41
		Smectic G	Plates 42–48, 75, 78
Smectic A_2	Plate 97	Smectic G'(J)	Plates 65, 66, 71, 80, 81
Smectic $\tilde{A}$	Plates 103, 104		
Smectic AB	Plates 13–15	Smectic H	Plates 51, 52
Smectic B	Plates 4–7, 10–12, 16, 17	Smectic H'(K)	Plates 49, 50, 66, 67, 81, 82
Smectic B (crystal)	Plates 8, 74, 76, 92, 93, 96	Smectic I	Plates 53, 54, 57, 63, 64, 79, 85
Smectic B (hexatic)	Plates 9, 88	Smectic I*	Plates 55–57, 70
		S_{VII} (TBBA)	Plate 58
Smectic B_2	Plates 16, 17	S_2 (CCH3)	Plate 59
Smectic C	Plates 18–23, 26, 61–63, 73, 84	S_3 (CCH3)	Plate 60
Smectic C*	Plates 23–25, 69		

Introduction

In recent years, experimental and theoretical studies of smectic liquid crystals have grown steadily in complexity as a widening range of polymorphic modifications of the smectic state has gradually been brought to light.

Even to the expert in the field, the situation concerning these many polymorphic modifications, their structure, and their nomenclature can be quite confusing. This arises partly because, with regard to structure, we are still in a position of developing knowledge and understanding, and partly because the history of the unfolding of the subject (and some misconceptions in the past) led, at one stage, to the use of a duality of nomenclature by different groups of workers in the field. To a comparative newcomer to the area, the literature on smectic polymorphism can be quite bewildering therefore; literature reports of the smectic properties of a given compound can employ not only different code letters to describe the same polymorphic smectic states, but also an even more confusing, inverted usage of pairs of code letters.

The positive identification of the polymorphic class to which a particular smectic modification belongs must nowadays rest on information gleaned from several experimental sources, including X-ray diffraction, but microscopic studies of the textures of the different smectic forms still remain, and will remain a powerful and economically practical experimental means of classifying smectic phases, particularly when these studies are combined with miscibility investigations.

The primary aim of this book has therefore been to present to the reader a collection of coloured photomicrographs of the textures of established examples of the different smectic polymorphic modifications. In choosing these photomicrographs, care has been taken to ensure that the textures are representative of the phase in question. That is, we have not yielded to the temptation to photograph abnormal regions of the texture arising from unusual deformations of the smectic structure caused by surface effects. In the belief that this book will be of service to the comparative beginner who is attempting to gain experience and knowledge in the field of the microscopy of smectics, as well as to those more expert in the subject, we have therefore been concerned to depict textures which are typical of the phase type and therefore likely to provide a good guide to the probable class of any unclassified smectic phases that may be encountered. In the case of the textures presented therefore, no special surface treatments of the glass microscope slides and coverslips have been used. Clean slides and coverslips, lightly wiped with lens tissue have been employed.

To ensure that pretransitional effects in the phases do not interfere, photomicrographs of textures have in all cases been taken with the temperature of the sample maintained well within the limits of the temperature range over which the smectic modification in question is thermodynamically stable. However, changes in texture that occur at a phase transition are frequently an important factor in phase identification; it was therefore judged useful to include some sequences of photomicrographs for specific compounds to illustrate the textural changes occurring as one phase type develops from another with change in temperature.

The text (Chapters 1 to 10) which accompanies the section on microscopic textures has been aimed at supplying to the reader a review of the sequence of development of knowledge relating to each of the smectic types. This has created one problem, since our understanding of smectic structure and indeed current views on whether particular modifications should be regarded as smectic liquid crystals or as crystal-type smectics (or even crystals) have been evolving during the writing process. We decided therefore that the text should carry an appendix (Chapter 10) in which parts of the text written earlier are updated in the light of current knowledge. To help to direct the reader's attention to the areas covered by Chapter 10, they are summarized below:

Hexatic and crystal smectic phases—with particular reference to B phases;
Antiphase behaviour and the role of bilayers in smectic A, B, and C phases;
Ferroelectric phases—chiral smectic C and other chiral tilted phases;
Structural features of smectic phases—a summarizing figure (page 153) and tables designed to function as *aides mémoires*.

It is hoped in particular that these chapters will provide an insight into the complex and often confusing literature on smectic liquid crystals. As mentioned already, aspects of nomenclature have been particularly problematical, and if these chapters remove some of the barriers which have in the past made if difficult for the increasing number of new workers involved in studies of smectic systems to divine and obtain a coherent view of smectic polymorphism, then they will have served a useful purpose. Since the text has been written for practically oriented readers whose backgrounds may be very varied (chemistry, physics, engineering, electronics, biology) because of the multidisciplinary involvements of liquid crystals, descriptions of smectic structure have been given in terms that will provide a simple mental picture. If this approach leads to some over-simplifications which offend those who are more theoretically or deeply physically oriented, we apologise and offer as the reason that we are attempting to guide the inexperienced and encourage newcomers to the field to become more involved in it. Details of understanding can be achieved once a secure basis has been established.

Since the main theme of the book concerns smectic liquid crystals and is aimed at an experimentally-involved readership, we did not want the written text to become disproportionate in length. Yet at the same time, we considered that it would be worthwhile to include in the section of photomicrographs pictures of representative textures of phases other than those of the smectic type. The object was to provide a more or less complete coverage of the textures of liquid crystals that may be encountered with mesogenic materials of the thermotropic type, concentrating upon the large and complex area of smectics, but providing coverage of typical textures of other phases and enabling visual comparisons of these with smectic textures to be made. The phases other than smectics to which we now refer are (a) nematic phases, (b) cholesteric phases, (c) discotic phases, and (d) 'blue' phases. We have chosen not to lengthen the written text by including chapters on each of these four categories of mesophase and there are additional reasons for this. Compared with our structural knowledge of smectics and the complexities of smectic polymorphism, nematic and cholesteric liquid crystals have been structurally well understood for many years, and neither type is complicated by polymorphism. They therefore seemed to require little introduction, and we have restricted ourselves to including at the end of this introduction a selection of key references to nematic and cholesteric phases—their structure, optical characteristics, etc.

A rather different situation obtains with categories (c) and (d). Speaking generally, *discotic phases* formed on heating compounds composed of relatively flat, disc-shaped molecules have only been recognized as a distinct class of liquid crystal for a few years. Knowledge concerning them is developing continuously and it becomes clear that not only do extended columnar discotic systems exist, but also discotic analogues of nematics are now well recognized. Polymorphism of the former type is possible, dependent upon the nature of the lateral packing of the columns and the tilt angle of the column with respect to the normal to the molecular planes. These and other fascinating aspects of discotic phases are slowly evolving. In this developing situation, a review of the field could only be up-to-date for a comparatively short period. Also, some excellent reviews already exist and others feature fairly regularly at conferences and symposia. Bearing these points in mind, and the fact that the compounds required to obtain textures representative of the increasing range of discotic modifications are not generally accessible, we have not written a chapter reviewing discotics, and have restricted ourselves to the inclusion of some photomicrographs representative in general terms of discotic systems. These are mainly concerned with providing the reader with a comparison with smectic textures. Again, at the end of this introduction we have given some key references to reviews on discotics written by workers active in the area.

By way of contrast, *'blue' phases* have been recognized for many, many years, but the narrow temperature range in which these phases exist between the upper thermal limit of the cholesteric phase and the formation of the isotropic liquid has limited progress on their study. Over the years, ideas relating to the molecular arrangement and structure of 'blue' phases have been proposed, and studies of the textures of 'blue' phases led us to suspect a long time ago (Coates and Gray, 1973, 1975) that at least two different 'blue' phase types could exist and that these were separated by a reversible, temperature-dependent transition. In the last three or four years, the 'blue' phase has come to attract much more attention, both from theoreticians and experimentalists. Here again therefore, we are dealing with a rather rapidly evolving situation, and we have decided not to attempt to review the current, shifting position regarding knowledge in this area. Instead, we append at the end of this introduction references to some of the recent research papers on the subject. The photomicrographs of 'blue' phases given in the book are therefore intended simply to show the reader what these phases look like microscopically in relation to the textures of other phases. Incidentally, it should be pointed out that one of the reasons why 'blue' phases escaped more general scrutiny at an earlier date is the fact that they exhibit little real texture when observed microscopically in transmission. Polarized microscopy must be used in the reflective mode if the rather beautiful mosaic and wrinkled mosaic textures illustrated in the section on photomicrographs are to be seen.

In our preface, we have included acknowledgments to members of the Liquid Crystal Group at the University of Halle, including the name of Professor Horst Sackmann. In our view, the introduction to this book would not be complete if we did not pay separate tribute to Professor Sackmann for his original pioneering work on the concept of miscibility as an indispensable aid to the identification of smectic phase type. Without this work and the phase classification system which he began, our knowledge of smectic liquid crystals would not be as clear as it is today.

Key References and Bibliography (in chronological order)

(a) and (b)   Nematic and Cholesteric Phases

Friedel, G. (1922). Les états mésomorphes de la matière. *Ann. Physique* **18**, 273.
De Vries, H.L. (1951). Rotatory power and other optical properties of liquid crystals. *Acta Cryst.* **4**, 219.

Brown, G.H., and Shaw, W.G. (1958). The mesomorphic state. Liquid crystals. *Chem. Rev.* **57**, 1049.
Maier, W., and Saupe, A. (1958–1960). Einfache molekularstatistische Theorie der nematischen Phase. *Z. Naturforsch.* (1958) **13a**, 564; (1959) **14a**, 882; (1960) **15a**, 287.
Gray, G.W. (1962). *Molecular Structure and the Properties of Liquid Crystals.* Academic Press, London and New York.
Ericksen, J.L. (1967). Continuum theory of liquid crystals. *Appl. Mech. Rev.* **20**, 1029.
Leslie, F.M. (1968). Thermal effects in cholesterics. *Proc. Roy. Soc.* **A307**, 359.
De Gennes, P.G. (1969). Long range ordering and thermal fluctuations in liquid crystals. *Mol. Cryst. Liq. Cryst.* **7**, 325.
Leslie, F.M. (1969). Continuum theory of cholesterics. *Mol. Cryst. Liq. Cryst.* **7**, 407.
Bouligand, Y. (1969–1974). A series of papers on optical properties, disclinations, and textures of cholesteric liquid crystals published in *J. Phys. (Paris).* (1969) **30**, 90; (1972) **33**, 525, 715; (1973) **34**, 603, 1011; (1974) **35**, 215, 959. Also (1973): Chevrons et quadrilatères dans les plages a éventails des cholestériques. *J. Microscop. (Paris)* **17**, 145.
De Gennes, P.G. (1971). Cristaux liquides nématiques. *J. Phys. (Paris)* **32**, 3.
Billard, J. (1972). Les phases mésomorphes et leur identification. *Bull. Soc. franç. Minéral.* **95**, 206.
Nehring, J., and Saupe, A. (1972). Schlieren texture in nematic and smectics. *J. Chem. Soc., Faraday Trans. II* **68**, 1.
Nehring, J. (1973). Calculations of the structure and energy of nematic threads. *Phys. Rev.* **A7**, 1737.
De Gennes, P.G. (1974). *The Physics of Liquid Crystals.* Clarendon Press, Oxford.
Demus, D., Demus, H., and Zaschke, H. (1974). *Flüssige Kristalle in Tabellen.* Deutscher Verlag für Grundstoffindustrie, Leipzig.
Gray, G.W., and Winsor, P.A. (eds.) (1974). *Liquid Crystals and Plastic Crystals,* Vol. I and II. Ellis Horwood Publishers, Chichester.
Bouligand, Y. (1975). Presentation of a film entitled: Textures of nematic and cholesteric phases. *J. Phys. (Paris)* **36**, 173.
Priestley, E.B., and Nojtowicz, P.J. (eds.) (1975). *Introduction to Liquid Crystals.* Plenum, New York.
Adamczyk, A., and Strugalski, Z. (1976). *Liquid Crystals.* W.N.T., Warsaw.
Chistyakov, I.G. (1976). *Liquid Crystals.* Ivanov. Gov. University, Ivanovo.
Elser, W., and Ennulat, R.D. (1976). Selective reflection in cholesteric liquid crystals. *In* G.H. Brown (ed.), *Advances in Liquid Crystals,* Vol. 2, Academic Press, London and New York, pp. 73–172.
Chandrasekhar, S. (1977). *Liquid Crystals.* Cambridge University Press, Cambridge.
Kléman, M. (1977). *Points, Lignes, Parois dans les Fluides Anisotropes et les Solides Cristallins,* Vol. 1 and 2. Les Editions de Physiques, Orsay.
Blinov, L.M. (1978). *Electro- and Magneto-optics of Liquid Crystals.* Nauka, Moscow.
Demus, D., and Richter, L. (1978). *Textures of Liquid Crystals.* Verlag Chemie, Weinheim.
Liebert, L. (ed.) (1978). *Liquid Crystals. Solid State Physics Supplement 14,* Academic Press, London and New York.
Luckhurst, G.R., and Gray, G.W. (eds.) (1979). *The Molecular Physics of Liquid Crystals.* Academic Press, London and New York.
De Jeu, W.H. (1980). *Physical Properties of Liquid Crystalline Materials.* Gordon and Breach Science Publishers, New York, London, and Paris.
Kelker, H., and Hatz, R. (1980). *Handbook of Liquid Crystals.* Verlag Chemie, Weinheim.

(c) Discotic Phases

An interesting example of theoretical predictions preceding experimental discovery was provided by discotic phases whose existence was foretold by de Gennes several years before Chandrasekhar produced practical examples of their occurrence in 1977.

Chandrasekhar, S., Sadashiva, B.K., and Suresh, K.A. (1977). Liquid crystals of disc-like molecules. *Pramana* **9**, 471.
Billard, J., Dubois, J.C., Tinh, N.H., and Zann, A. (1978). Une mesophase discotique. *Nouveau J. de Chemie* **2**, 535.

Béguin, A., Billard, J., Dubois, J.C., Tinh, N.H., and Zann, A. (1979). Discotic mesophase potentialities. *J. Phys. (Paris)* **40**, 15.
Cotrait, M., and Marsau, P. (1979). Crystalline arrangement of some disc-like compounds. *J. Phys. (Paris)* **40**, 519.
Destrade, C., Mondon-Bernaud, M.C., and Tinh, N.H. (1979). Mesomorphic polymorphism in some disc-like compounds. *Mol. Cryst. Liq. Cryst.* **49**, *169.*
Destrade, C., Mondon, M.C., and Malthête, J. (1979). Hexasubstituted triphenylenes: a new mesomorphic order. *J. Phys. (Paris)* **40**, 17.
Levelut, A.-M. (1979). Structure of a disc-like mesophase. *J. Phys. (Paris)* **40**, L81.
Tinh, N.H., Destrade, C., and Gasparoux, H. (1979). Nematic disc-like liquid crystals. *Phys. Lett.* **72A**, 251.
Billard, J. (1980). Discotic mesophases. A review. In W. Helfrich and G. Heppke (eds.), *Liquid Crystals of One- and Two-Dimensional Order, Springer Series in Chemical Physics 11,* Springer, Berlin, Heidelberg, and New York, pp. 383–395.
Destrade, C., Bernaud, M.C., Gasparoux, H., Levelut, A.-M., and Tinh, N.H. (1980). Disc-like mesogens with columnar and nematic phases. In S. Chandrasekhar (ed.), *Liquid Crystals,* Heyden and Son, London, pp. 29–33.
Destrade, C., Malthête, J., Tinh, N.H., and Gasparoux, H. (1980). Truxene derivatives: temperature inverted nematic-columnar sequence in disc-like mesogens. *Phys. Lett.* **78A**, 82.
Gasparoux, H. (1980). Carbonaceous mesophase and disc-like molecules. In W. Helfrich and G. Heppke (eds.), *Liquid Crystals of One- and Two-Dimensional Order, Springer Series in Chemical Physics 11,* Springer, Berlin, Heidelberg, and New York, pp. 373–382.
Gasparoux, H., Destrade, C., and Fug, G. (1980). Carbonaceous mesophase and disc-like liquid crystals. *Mol. Cryst. Liq. Cryst.* **59**, 109.
Levelut, A.-M. (1980). X-ray diffraction by mesophases of some hexa-alkanoates of terphenylene. In S. Chandrasekhar (ed.), *Liquid Crystals,* Heyden and Son, London, pp. 21–27.
Sigaud, G., Achard, M.F., Destrade, C., Tinh, N.H. (1980). In W. Helfrich and G. Heppke (eds.), *Liquid Crystals of One- and Two-Dimensional Order, Springer Series in Chemical Physics 11,* Springer, Berlin, Heidelberg, and New York, pp. 403–408.
Billard, J., Dubois, J.C., Levelut, A.-M., and Vauchier, C. (1981). Structure of the two discophases of rufigallol hexa-n-octanoate. *Mol. Cryst. Liq. Cryst.* **66**, 115.
Brand, H., and Pleiner, H. (1981). Hydrodynamics of biaxial discotics. *Phys. Rev. A* **24**, 2777.
Chandrasekhar, S. (1981). Liquid crystals of disc-like molecules. *Mol. Cryst. Liq. Cryst.* **63**, 171.
Destrade, C., Gasparoux, H., Babeau, A., and Tinh, N.H. (1981). Truxene derivatives: a new family of disc-like liquid crystals with an inverted nematic-columnar sequence. *Mol. Cryst. Liq. Cryst.* **67**, 37.
Destrade, C., Tinh, N.H., Gasparoux, H., Malthête, J., and Levelut, A.-M. (1981). Disc-like mesogens. *Mol. Cryst. Liq. Cryst.* **71**, 111.
Destrade, C., Tinh, N.H., Malthête, J., and Jacques, J. (1981). On a 'cholesteric' phase in disc-like mesogens. *Phys. Lett.* **79A**, 189.
Gasparoux, H. (1981). Carbonaceous mesophase and disc-like nematic liquid crystals. *Mol. Cryst. Liq. Cryst.* **63**, 231.
Otani, S. (1981). Carbonaceous mesophase and carbon fibers. *Mol. Cryst. Liq. Cryst.* **63**, 249.
Tinh, N.H., Bernaud, M.C., Sigaud, G., and Destrade, C. (1981). Disymmetric hexasubstituted triphenylenes. *Mol. Cryst. Liq. Cryst.* **65**, 307.
Vauchier, C., Zann, A., Le Barny, P., Dubois, J.C., and Billard, J. (1981). Orientation of discotic mesophases. *Mol. Cryst. Liq. Cryst.* **66**, 103.
Bunning, J.D., Lydon, J.E., Eaborn, C., Jackson, P.M., Goodby, J.W., and Gray, G.W. (1982). Classification of the mesophase of di-isobutylsilanediol. *J. Chem. Soc., Faraday Trans. I* **78**, 713.
Chandrasekhar, S. (1982). Liquid crystals of disc-like molecules. In G.H. Brown (ed.), *Advances in Liquid Crystals,* Vol. 5. Academic Press, London and New York, pp. 47–78.
Chandrasekhar, S. (1983). Liquid crystals of disc-like molecules. *Proc. Roy. Soc.* **A309**, 93.
Destrade, C., Foucher, P., Gasparoux, H., Tinh, N.H., Levelut, A.-M., and Malthête, J.

(1983). Disc-like mesogen polymorphism. Presented at the Ninth International Conference on Liquid Crystals, Bangalore, India (1982)—to appear in *Mol. Cryst. Liq. Cryst.*

Godquin-Giroud, A.-M., and Billard, J. (1983). Thermotropic transition metal complex discogens and their smectogenic intermediates. *Mol. Cryst. Liq. Cryst.* **97**, 287.

Guillon, D., Skoulios, A., Piechocki, C., Simon, J., and Weber, P. (1983). Discotic mesophases of the metal-free derivative of octa(dodecyloxymethyl)phthalocyanine. *Mol. Cryst. Liq. Cryst.* **100**, 275.

Lejay, L., and Pesquer, M. (1983). A theoretical study of mesogenic disc-like compounds: the hexa-n-alkoxy- and hexa-n-alkanoyloxy-truxene series. *Mol. Cryst. Liq. Cryst.* **95**, 31.

(d) 'Blue' Phases

Though currently attracting considerable attention, the 'blue' phase region which precedes formation of the cholesteric phase on cooling the isotropic liquid phase of a cholesterogen was probably recognized in the writings of Reinitzer and Lehmann in the period 1888–1889.

Gray, G.W. (1956). The mesomorphic behaviour of the fatty esters of cholesterol. *J. Chem. Soc.* 3733.

Saupe, A. (1969). On molecular structure and physical properties of thermotropic liquid crystals. *Mol. Cryst. Liq. Cryst.* **7**, 59.

Coates, D., and Gray, G.W. (1973). Optical studies of the amorphous liquid-cholesteric liquid crystal transition: the Blue phase. *Phys. Lett.* **45A**, 115.

Gray, G.W., and Winsor, P.A. (1974). In G.W. Gray and P.A. Winsor (eds.), *Liquid Crystals and Plastic Crystals*, Vol. 1, Ellis Horwood Publishers, Chichester, p. 1 and 16.

Coates, D., and Gray, G.W. (1975). A correlation of optical features of amorphous liquid-cholesteric liquid crystal transitions. *Phys. Lett.* **51A**, 335.

Bergmann, K., and Stegemeyer, H. (1979). Evidence for polymorphism within the so-called 'Blue Phase' of cholesteric esters. Part I. Calorimetric and microscopic measurements. *Z. Naturforsch.* **34A**, 251. Part II. Selective reflection and optical rotatory dispersion. *Z. Naturforsch.* **34A**, 253. Part III. The circular dichroism of the blue phase at high pressures. *Z. Naturforsch.* **34A**, 255. Part IV. The temperature and angular dependence of selective reflection. *Z. Naturforsch.* **34A**, 1033.

Hornreich, R.M., and Shtrikman, S. (1980). Theory of BCC orientational order in chiral liquids: the cholesteric blue phase. In W. Helfrich and G. Heppke (eds.), *Liquid Crystals of One- and Two-Dimensional Order, Springer Series in Chemical Physics 11*, Springer, Berlin, Heidelberg, and New York, pp. 185–195.

Meiboom, S., and Sammon, M.J. (1980). Structure of the blue phase of a cholesteric liquid crystal. *Phys. Rev. Lett.* **44**, 882.

Stegemeyer, H., and Bergmann, K. (1980). Experimental results and problems concerning 'Blue Phases'. In W. Helfrich and G. Heppke (eds.), *Liquid Crystals of One- and Two-Dimensional Order, Springer Series in Chemical Physics 11*, Springer, Berlin, Heidelberg, and New York, pp. 161–175.

Finn, P.L., and Cladis, P.E. (1981). The cholesteric-isotropic—blue phase triple point. *Mol. Cryst. Liq. Cryst. Lett.* **72**, 107.

Marcus, M.A., and Goodby, J.W. (1981). Cholesteric pitch and blue phases in a chiral-racemic mixture. *Mol. Cryst. Liq. Cryst. Lett.* **72**, 297.

Marcus, M.A. (1981). Quasicrystalline behaviour and phase transition in cholesteric 'blue' phase. *J. Phys. (Paris)* **42**, 61.

Meiboom, S., and Sammon, M.J. (1981). Blue phases of cholesteryl nonanoate. *Phys. Rev.* **A24**, 468.

Meiboom, S., Sethna, J.P., Anderson, P.W., and Brinkman, W.F. (1981). Theory of the blue phase of cholesteric liquid crystals. *Phys. Rev. Lett.* **46**, 1216.

Finn, P.L., and Cladis, P.E. (1982). Cholesteric blue phases in mixtures in an electric field. *Mol. Cryst. Liq. Cryst.* **84**, 159.

Onusseit, H., and Stegemeyer, H. (1982). Investigation of the blue phase in a compensable cholesteric mixed system. *Chem. Phys. Lett.* **89**, 95.

Sammon, M.J. (1982). Numerical three dimensional relaxation of liquid crystal director fields. *Mol. Cryst. Liq. Cryst.* **89**, 305.

Bensimon, D., Domany, E., and Shtrikman, S. (1983). Optical activity of cholesteric liquid crystals in the pretransitional regime and in the blue phase. *Phys. Rev.* **28A,** 427.

Crooker, P.P. (1983). The cholesteric blue phase: a progress report. *Mol. Cryst. Liq. Cryst.* **98,** 31.

Grebel, H., Hornreich, R.M., and Shtrikman, S. (1983). Landau theory of cholesteric blue phases. *Phys. Rev.* **28A,** 1114.

Grebel, H., Hornreich, R.M., and Shtrikman, S. (1983). Theoretical NMR spectra of cubic cholesteric blue phases. *Phys. Rev.* **28A,** 2544.

Hornreich, R.M., and Shtrikman, S. (1983). Theory of light scattering in cholesteric blue phases. *Phys. Rev.* **28A,** 1791.

Meiboom, S., Sammon, M.J., and Brinkman, W.F. (1983). Lattice of disclinations: the structure of the blue phases of cholesteric liquid crystals. *Phys. Rev.* **27A,** 438.

Onusseit, H., and Stegemeyer, H. (1983). Induced blue phases in cholesteric mixed systems. *Chem. Phys. Lett.* **94,** 417.

Onusseit, H., and Stegemeyer, H. (1983). Growth of cubic liquid single crystals in cholesteric blue phases. *J. Cryst. Growth* **61,** 409.

Berreman, D.W. (1984). In A.C. Griffin and J.F. Johnson (eds.). *Liquid Crystals and Ordered Fluids,* Vol. 4, Plenum Press, New York, to be published.

Clark, N.A., Vohra Sandeep, T., and Handschy, M.A. (1984). Elastic resonance of a liquid-crystal blue phase. *Phys. Rev. Lett.* **52,** 57.

Marcus, M.A. (1984). In A.C. Griffin and J.F. Johnson (eds.). *Liquid Crystals and Ordered Fluids,* Vol. 4, Plenum Press, New York, to be published.

1 The smectic A phase

Introduction

The smectic A phase is the smectic polymorphic modification which possesses least order. In any phase sequence which includes smectic A, this phase precedes all other smectic phases upon cooling either the isotropic liquid or the nematic phase.

The phase was first studied in detail by Friedel (1922) who reported his observations in his classical article in *Annales de Physique*. In fact, Friedel examined only smectogens of the A type, and smectic polymorphism lay undetected for many years. Therefore, the smectic A phase was not assigned a code letter until much later by Sackmann and Demus (1966).

Grandjean (1917) first made preliminary investigations into the structure of what is now known as the smectic A phase. He found that uncovered droplets of a smectic A, supported on mica or very clean glass and viewed microscopically, showed a number of steps along their edges; these steps became known as Grandjean terraces. The steps were of unequal thickness, but their heights always corresponded to multiples of the molecular length of the smectogen. Friedel (1922) considered that these stepped drops provided a close analogy to the stepped drops formed on evaporation of soap solutions. These results suggested that the mesophase consisted of a stratified molecular arrangement, a conclusion which is in fact valid for all the smectic polymorphic modifications except smectic D.

Structure of the smectic A phase

Early X-ray diffraction studies of the smectic A phase had shown that the molecules are arranged with their molecular long axes perpendicular to the planes of the layers. The lateral distribution of the molecules within each layer is however random, and the molecules in the smectic A layers are able to rotate freely about their long axes. If we consider that the molecular structure of the mesogen is flat and oblong in shape, i.e.

blade-like or lath-like, and that rotation about the molecular long axis will sweep out a volume which is cylindrical, then the structure of a smectic A layer would be of the form shown in Fig. 1.1. Although this simple model is

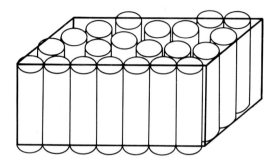

Figure 1.1 Disordered arrangement of molecular centres in a smectic A phase.

acceptable for many purposes, a number of flaws have been discovered through more detailed investigations made in recent years. The various points to which attention must be paid are as follows:

(a) Numerous X-ray studies have shown that the lamellar spacings (d) for various smectic A phases are considerably shorter than the corresponding molecular lengths (l) calculated for the molecules in their fully extended, all-*trans* conformations and/or configurations (e.g. $d:l$ may be 0.8:1).

(b) In some cases, however, notably the 4-alkyl-4'-cyanobiphenyls (Leadbetter, Durrant, and Rugman, 1977; Leadbetter and co-workers, 1979a) and related materials, the lamellar spacings (d) are larger than the calculated molecular lengths (l); the ratio is of the order of 1.4:1.

(c) Quasielastic neutron scattering experiments showed only small differences between the results with the scattering vector probing the molecular motions perpendicular or parallel to the lamellar planes (Leadbetter and co-workers, 1975a, b, 1976, 1976b; Bonsor and co-workers, 1978; Dianoux and co-workers, 1976).

As most mesogens that exhibit the smectic A phase have molecular structures that consist of a rigid central core plus one or two terminal alkyl chains per core, then the shorter lamellar spacings were initially accounted for either by liquid-like alkyl chains or by random tilting of the core plus chain deformations. The situation for a liquid-like distribution of the terminal chains is shown in Fig. 1.2(a), and for core-tilting with terminal chain deformation in Fig. 1.2(b).

Secondly, lamellar spacings greater than the calculated molecular lengths were interpreted in terms of bilayer structures. These (interdigitated) structures will be discussed separately—see also Chapter 10.

THE SMECTIC A PHASE

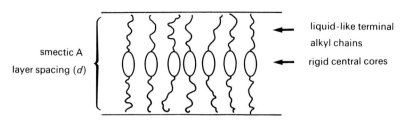

Figure 1.2(a) Terminal chain compression in the smectic A phase (structure stressed).

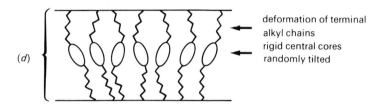

Figure 1.2(b) Terminal chain deformation and core tilting in the smectic A phase; the average alignment of the molecules with respect to the layer planes would be orthogonal (uniaxial).

Thirdly, the results of the neutron scattering experiments seem to imply that the disposition of the molecules in each layer is identical with respect to both the layer plane and the normal to the layer plane. One possible interpretation of this *might* be that the molecular long axes are tilted at an angle of 45° to the layer planes. Such an arrangement is shown in Fig. 1.3.

For mesogens that have lamellar spacings shorter than the overall molecular length, the suggestion of deformed or tilted structures (Fig. 1.2(a) and (b)) or structures with large tilt angles (Fig. 3) can be discounted for the following reasons.

(a) The particular layer spacing adopted by a smectic A is a consequence of the structure(s) of the individual molecular components, and not

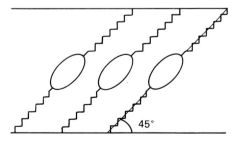

Figure 1.3 At first sight, the molecules would have to have large tilt angles with respect to the S_A layers in order to comply with neutron scattering data.

some predetermined value. It would be difficult to justify the adoption of a *smaller* layer spacing by the smectic A phase, as a result of which the molecules within the layers and the layers themselves would be put under stress.

(b) A large tilt angle of 45° would give the phase a birefringent texture. However, the phase exhibits a positive uniaxial interference figure by conoscopic observation, and also a homeotropic microscopic texture is adopted when the layers lie flat on the supporting surfaces. Moreover, a tilt angle of 45° would give lamellar spacings much smaller in relation to the calculated molecular lengths than those that are obtained experimentally.

Lösche (1973, 1974) tried to overcome some of these structural problems by suggesting that the molecules gyrate about their centres of mass, giving a small, time-dependent tilt angle of the order of 10–20°. This model is shown in Fig. 1.4.

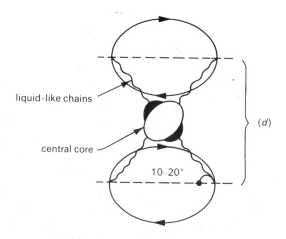

Figure 1.4 Gyroscopic model of the smectic A phase.

This model would, however, require that all the molecules in a particular domain should move together in harmony, in order to retain the continuity of the layer structure. Moreover, the cylindrical volume occupied by one molecule in gyroscopic motion would be much greater than that occupied by the same molecule rotating about its long axis in the cylindrical model.

Modern theories of the smectic A phase now propose that the molecules are randomly tilted within the individual layers, the tilt angle varying from 10–25°. Since the tilt direction is random, the situation would rapidly average out over a number of layers to give a mean orthogonal, uniaxial arrangement of the molecular long axes. This is unlike the situation for the tilted smectic C phase, in which the molecules in one domain have the same preferred tilt direction.

The observations made in the neutron scattering experiments are now accounted for by small up and down oscillations of the molecules within the layers accompanied by rotation of the molecules about their molecular long axes (Leadbetter and Richardson, 1979b, and references cited therein).

Leadbetter (1979c), and Leadbetter and Richardson (1979b) have also suggested that the order parameter of the phase should be involved when producing a structural model for the phase. De Vries, Ekachai, and Spielberg (1979) have developed this idea even further by taking into account the expansions of the molecules which will occur at the elevated temperatures involved in most measurements of lamellar spacings. They postulated that for smectic A phases in general, the long axes of neighbouring molecules are approximately parallel to one another. The degree of parallel order is given by the orientational order parameter.

$$S = \frac{1}{2}\langle 3\cos^2\theta - 1\rangle$$

where θ is the angle between the long axis of a given molecule and the average direction of orientation of the molecular long axes (the director) for the population of molecules. Although the experimental values for S vary considerably from material to material, an average value would be about 0.8. De Vries, Ekachai, and Spielberg (1979) have shown that for an average value of $S=0.8$, a probability distribution for the tilt angle θ gives $\theta_{max}=14.2°$ and $\langle\theta\rangle=19.3°$. Therefore, the average tilt angle in the smectic A phase is quite large. They went on to make allowances for thermal expansion of the molecules, and by doing this they were able to calculate values of S and compare them with experimental values. Their conclusion was that the observed and calculated values agree fairly well, and that there should therefore be no deformation or twisting of the molecules within the layers. The model for the smectic A phase obtained by this treatment is best summarized in the words of de Vries and his co-workers (1979). 'The molecules are extended, with no appreciable interpenetration or 'kinking'. The average direction of the long (molecular) axes is perpendicular to the smectic planes, but the long axes of individual molecules can make quite large tilt angles with the plane normal; . . . This tilt is a simple consequence of the imperfect parallel alignment of the molecules inherent to the smectic A phase. There is no long-range order in the direction of tilt. The layer thickness is not constant, but fluctuates as a function of the local average tilt angle.' This layer structure is shown in Fig. 1.5 which attempts to depict the very diffuse layer situation. Density fluctuation plots suggest that the layers only just exist, and this must be borne in mind in the section on textures where, for convenience, the discussion is developed in terms of finite layers.

Figure 1.5 The diffuse layer structure of the smectic A phase.

The bilayer smectic A phase

All those smectogens of the A type that are known to have bilayer structures appear to have molecular structures that incorporate a terminal cyano group. In the case of the 4-alkyl- and 4-alkoxy-4'-cyanobiphenyls, the average lamellar spacing is about 1.4 times the molecular length (Leadbetter, Durrant, and Rugman, 1977; Leadbetter and co-workers, 1979a), but other values intermediate between 1 and 2 occur in other cyano-systems (Leadbetter, 1980). With such ratios, this suggests that the layer structure is of an interdigitated bilayer type. A variety of packing arrangements of this type may be envisaged for the cyanobiphenyls, as shown in Fig. 1.6.

Leadbetter (1975a, 1977, 1979a) and Bonsor (1978) and their co-workers have made extensive studies of these types of system by X-ray diffraction and by neutron scattering experiments involving perdeuteriated and selectively deuteriated mesogens. It was concluded that the materials have bilayer smectic A packing arrangements of type (a) (Fig. 1.6). Considering various possible packing arrangements and core-core associations, this layer structure is the most logical of the three. Type (c) would give repulsive interactions between adjacent cyano-groups and also between adjacent layers, if there were an alternating packing of the layers. Type (b) would give a shorter bilayer spacing than that observed, along with repulsive interactions between layers. There is also strong evidence (Leadbetter, Richardson, and Colling, 1975b) to support the view that this interdigitated arrangement persists in small aggregates of molecules in the nematic phases produced on heating such smectic A phases, and even in the isotropic liquids. This type of bilayer arrangement is also known to

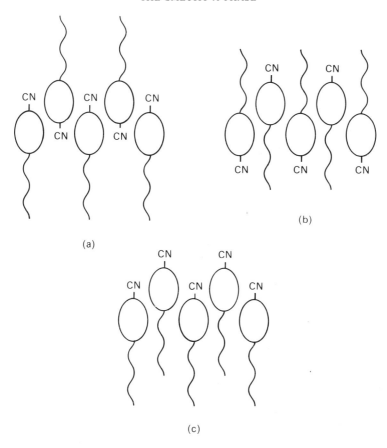

Figure 1.6(*a*), (*b*), (*c*) Possible bilayer structures for the interdigitated bilayer smectic A phase of 4-alkyl- or 4-alkoxy-4'-cyanobiphenyls.

occur for the nematic phases of *trans*-1-alkyl-4-(4'-cyanophenyl)cyclo hexanes and 1-alkyl-4-(4'-cyanophenyl)bicyclo [2,2,2,] octanes (Leadbetter, 1980).

The dielectric anisotropies ($\triangle_\varepsilon$) for materials of this type are quite large, e.g. for the nematic phase of 4-cyano-4'-pentylbiphenyl, $\varepsilon_\parallel - \varepsilon_\perp \sim 11$ (Gray, Harrison, and Nash, 1973). However, experiments made with such nematic phases diluted with other nematogens of weakly positive or even negative dielectric anisotropy show that $\triangle_\varepsilon$ *increases*. The diluent molecules break-up the molecular pairing (which minimizes $\triangle_\varepsilon$) and extrapolation of the results shows that if the pure material did *not* have a bilayer packing, then the 'molecular' dielectric anisotropy would be of the order 35. This gives very good supporting evidence for a bilayer structure incorporating opposed dipoles.

The existence of re-entrant nematic phases (Cladis and co-workers, 1975, 1978; Madhusudana, Sadashiva, and Moodithaya, 1979) for materials of this type also gives supporting evidence for the interdigitated bilayer structure of their smectic A phases.

Textures of the smectic A phase

There are only two important microscopic textures exhibited by the smectic A phase—the homeotropic or pseudoisotropic texture and the focal-conic fan texture.

The homeotropic or pseudoisotropic texture

When the homeotropic texture is observed using a polarizing microscope, and crossed polarizers, it appears black, except in the neighbourhood of deformations, e.g. around air bubbles or particles of impurity. These areas are often ringed by bright birefringent regions due to structural dislocations at the edge of the bubble or impurity.

The homeotropic texture can be prepared by using a very clean glass slide and coverslip. Clean surfaces are obtained by soaking the glass in concentrated nitric acid for an hour, washing with distilled water, and finally rinsing with acetone or propan-2-ol (IPA) which is then allowed to drain off and evaporate. Thin preparations can also induce homeotropy; this can be achieved by melting the material on a clean glass slide and allowing the isotropic liquid to run under the coverslip by capillary action. Homeotropic textures can also be produced with the aid of surface treatment by surfactants, e.g. lipids such as lecithin, and materials such as substituted cinnamic acids.

In the homeotropic texture, the molecules are aligned with an average perpendicular orientation (Fig. 1.5) with respect to the layers which lie parallel to the surfaces (Fig. 1.7). It is important in experiments aimed at the identification of a sequence of smectic phases to obtain this texture for the smectic A phase, particularly if other phases formed on cooling have tilted structures.

The focal-conic fan texture of the smectic A phase

The focal-conic fan texture is probably the most commonly observed (Friedel, 1922) texture of the smectic A phase (see Plate 1). The phase usually separates out (on cooling the isotropic liquid or the nematic phase) in the form of bâtonnets (see Plate 2). The bâtonnets then coalesce and build up the focal-conic texture. The focal-conic fan texture contains optical discontinuities which are visible in ordinary light and with crossed

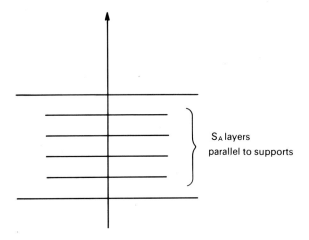

Figure 1.7 Homeotropic texture of the smectic A phase.

polarizers, and which appear as dark lines in the form of ellipses and hyperbolae. Each ellipse is associated with one hyperbola, and the pair of lines are related as a focal-conic pair. Such pairs of focal-conic lines are a direct consequence of particular arrangements of concentric, equidistant layers. Sections through the layers (*a*) in the plane of the ellipse give concentric circles, and (*b*) in the plane of the hyperbola give two sets of concentric circles, such that individual circles are paired-off in the two sets. The sets of circles are known as Dupin cyclides.

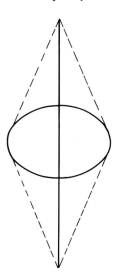

Figure 1.8 The simple straight line and circle relationship for a cyclide.

To simplify the treatment of the structure giving rise to the focal-conic fan texture, consider the special situation in which the pair of focal-conics (usually an ellipse and hyperbola) consists of a circle and straight line (Fig. 1.8). The enclosed focal-conic domain consists of two cones with a common circular base. The set of parallel, curved surfaces associated with this situation resembles a doughnut or the shape of a motor-car tyre (see Fig. 1.9). Any cross-section (containing the straight line of Fig. 1.8) through this arrangement shows the equally-spaced layers of molecules as concentric circles, e.g. circle A' and circle B' in the upper part of Fig. 1.9. The molecules are of course arranged perpendicular to the layers in the cyclide. Therefore the molecules in the plane of any such section, although restricted to their layers, appear to radiate from the centres E and F of the circles.

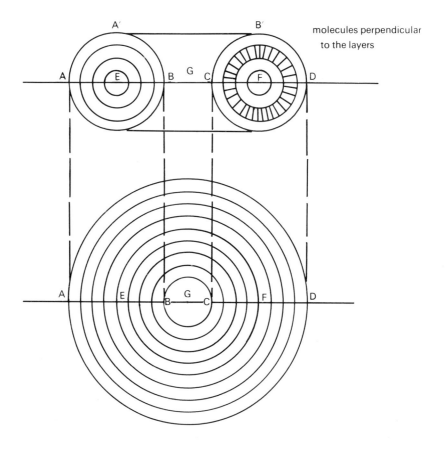

Figure 1.9 Cross-section and plan of the 'Dupin' cyclides for the circle and straight line.

THE SMECTIC A PHASE

In order to fill the space available in an efficient way, layers of molecules can be added, thus increasing the diameters AB and CD in Fig. 1.9. Eventually B and C may become coincident, i.e. the two cross-sectional circles will touch at point G. At this stage the central hole will no longer exist and will be replaced by the point G. As further layers are added, they must meet vertically above and below G. These intersections of the layers will build up a number of points in the plane of the page in the upper half of Fig. 1.9, such that they form a straight line passing through G. Figure 1.10 shows the situation, i.e. as a development of the top half of Fig. 1.9.

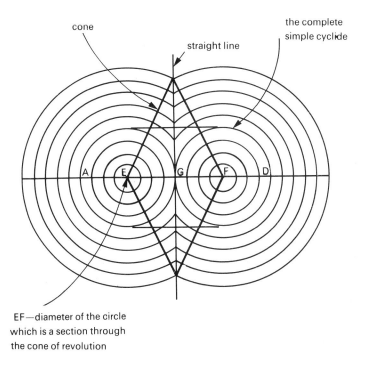

Figure 1.10 A vertical cross-section through the set of cyclides.

The straight line formed in this way is the locus of the apices of cones of revolution of which the original circle (Fig. 1.8) and the circle of diameter EGF in the lower part of Fig. 1.9 is a common section. In this way the structure appears to obtain its lowest form of potential energy. A set of full Dupin cyclides is never actually observed, and in this texture of the S_A phase we have just a continuum of focal-conic domains. In the simple case under discussion, the heavily lined area in Fig. 1.10 is a section through the focal-conic domain consisting of two cones with a common circular base.

The focal-conic domain (and conic sections through it) for the straight line and circle relationship is shown in Fig. 1.11(a). This depicts the complete focal-conic domain and the positions of the molecules in this arrangement. Figure 1.11(b) represents the shape of one particular molecular layer in the focal-conic domain of Fig. 1.11(a).

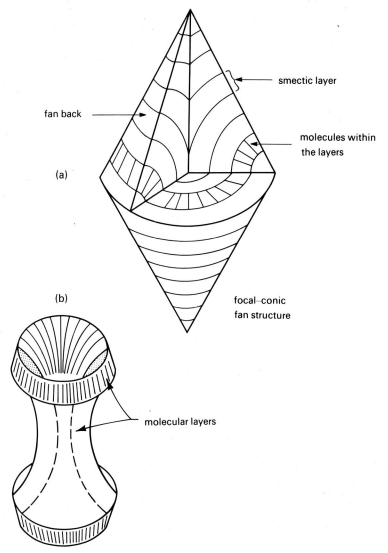

Figure 1.11(a) Focal-conic domain for the special case of ellipse=circle, and hyperbola=straight line. The short, straight lines represent molecules.
(b) A diagram of one molecular layer in the simple focal-conic domain. Redrawn from Coates and Gray, *The Microscope* (1977).

If we now substitute an ellipse for the circle and a hyperbola for the straight line, and retain the concentric packing arrangement of the layers of molecules, then we have the true or commonly observed focal-conic structure. The arrangement of Dupin cyclides is shown in Fig. 1.12. Again the sets of surfaces form a doughnut shape, but the hole (QR) is off-centre compared with that for the simple situation in Fig. 1.9, and one side (PQ) of the doughnut is more enlarged than the other. The molecular layers form concentric tubes about the centre of the hole (QR) as shown in the upper part of Fig. 1.12. The lower part of Fig. 1.12 shows the cross-section through PQRS. Note that we no longer have a simple set of concentric circles, and that crescent-shaped regions now exist.

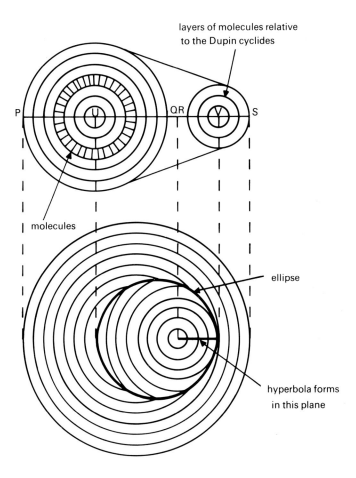

Figure 1.12 Cross-section and plan of the Dupin cyclides for the case of an ellipse and a hyperbola. Based on Hartshorne and Stuart, *Crystals and the Polarising Microscope* (1970).

As in the simple straight line and circle relationship, other layers may now be added until the hole is completely filled and becomes a point (T). As further layers are added they meet above and below the point T (see Fig. 1.13). These points of intersection form the hyperbola which has the

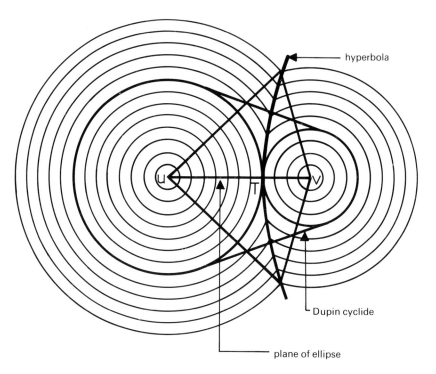

Figure 1.13 A vertical cross-section through the set of cyclides. Based on Hartshorne and Stuart, *Crystals and the Polarising Microscope* (1970).

same locus as the ellipse which passes through U and V. As in the simple cyclide structure, the *hyperbola* forms the *locus* of the *apices* of *cones* of *revolution* of which the *ellipse* is a *common section*. Again, only the focal-conic domain, and not the whole set of Dupin cyclides, is obtained, and this structure appears to represent the lowest potential energy for the packing arrangement of the molecules in liquid-like layers.

The ellipse and hyperbola have a definite mathematical relationship and they are termed a 'pair' of focal-conics. For the ellipse and hyperbola we have the relationships (after Hartshorne and Stuart, 1970):

$$\frac{x^2}{a^2} + \frac{y^2}{b^2} = 1 \quad \text{for the ellipse}$$

and
$$\frac{x^2}{a'^2} - \frac{y^2}{b'^2} = 1 \quad \text{for the hyperbola}$$

and it can be shown that $a' = \sqrt{(a^2 - b^2)}$

and $b' = b$

In a similar fashion to that for the straight line and circle, we can represent a perfect or full focal-conic domain, and sections through it, as shown in Fig. 1.14.

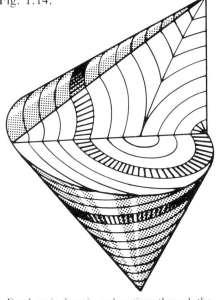

Figure 1.14 Focal-conic domain and sections through the domain for the case of an ellipse and hyperbola.

The existence of the focal-conic relationship can be confirmed by detailed microscopic observation of the texture. The ellipses and hyperbolae of the focal-conic domains appear as dark intersecting lines, i.e. as optical discontinuities, because they mark sharp local changes in direction of the optic axis. The fan-shaped areas are the surfaces of the focal-conic domains flattened on the glass supports.

However, it is rare that a perfectly formed focal-conic domain is observed. A number of deformed focal-conics may be obtained and some are shown in Fig. 1.15.

Finally, it should be noted that these naturally occurring focal-conic domains do not have *perfectly* related ellipses and hyperbolae. The lines are either not exactly confocal or not true conics; in fact some theoretical considerations suggest that not all the layers can be arranged in Dupin cyclides and that there must be additional layers.

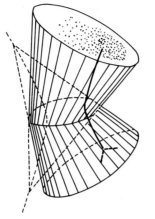

(b) Half a domain with one branch of hyperbola.

(a) Two imperfect domains sharing a common ellipse.

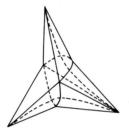

(d) Domain with strongly reduced conics.

(c) Twin domain that has not been fully developed.

Figure 1.15 Deformed focal-conics (after Friedel and Grandjean (1910)).

Focal-conic textures occur for all the other smectic polymorphic modifications except for the smectic D phase, and further reading on focal-conic textures and their structural implications is to be found in the article of Friedel (1922), in the book of Hartshorne and Stuart (1970), and in the article of Bouligand (1972).

The polygonal texture of the smectic A phase

This texture (Friedel, 1922) is a special form of the focal-conic fan texture; Plate 3 shows the polygonal texture of n-butyl 4-(4'-phenylbenzyl-ideneamino)cinnamate. The texture is only exhibited when a thick-film preparation is used; its occurrence in the perfect form of Plate 3 is quite exceptional, and a thick preparation is no guarantee that it will be obtained.

The polygonal texture is made up of focal-conic domains which are not full domains, but in fact half domains; that is, one branch of the hyperbola,

THE SMECTIC A PHASE

either above or below the plane of the ellipse is missing, and therefore the domain takes on a conical shape. These cones lie with their elliptical bases supported on one glass surface, and their apices touching the other parallel surface. Similar cones lie with their elliptical bases on the opposite surface and their apices touching the other support. Thus, the space available is filled efficiently (Fig. 1.16). These two interpenetrating sets of cones are

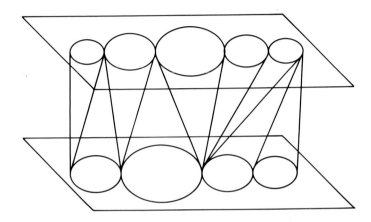

Figure 1.16 The focal-conic domains filling space efficiently.

not randomly packed however; they are packed together in families such that the ellipses of one family are tangential to one another and to the sides of a polygonal area which contains them; the polygonal area (see Fig. 1.17) may have numerous sides. In Plate 3, these polygonal areas can be seen

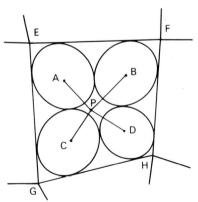

Figure 1.17 A polygonal area on the upper supporting surface. Redrawn from Hartshorne and Stuart, *Crystals and the Polarising Microscope* (1970).

    A to D: foci of ellipses
    AP, BP, etc: projections of hyperpolae onto upper surface
         ≡ major axes of the ellipses extended to meet at P.

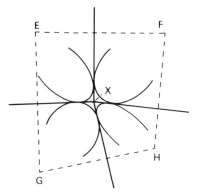

Figure 1.18 A polygonal domain showing the point X at which the hyperbolae meet on the lower surface; the lines from X are portions of sides of a corresponding set of polygonal areas on the lower surface. Redrawn from Hartshorne and Stuart, *Crystals and the Polarising Microscope* (1970).

defined by dark lines, which are in fact slightly curved. Note how the elliptical sections within these areas just touch the sides of each polygon. Smaller ellipses naturally fill the spaces between the large ellipses in Fig. 1.17 (and Fig. 1.18).

Furthermore, the hyperbolae belonging to ellipses in the same polygonal area (say on the upper surface) all meet on the lower surface at one point X (see Fig. 1.18) which is immediately below the point of intersection P of the major axes of the ellipses on the upper surface. This point of intersection of the hyperbolae is in fact one corner of several polygonal areas on the lower surface. The polygonal edges that emanate from this point on the lower surface intersect the projections (onto the lower surface) of the sides of the polygon on the upper surface at right angles.

In the polygonal texture, the elliptical sections observed by optical microscopy always show straight brushes crossing the section in the same direction for all the ellipses.

A section through a thick polygonal film is shown in Fig. 1.19; for

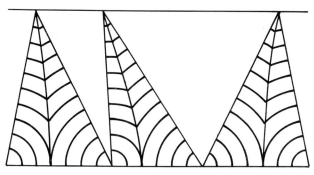

Figure 1.19 Continuity of layers within the S_A phase.

simplicity, the curved layers have been drawn only in sections through cones with their elliptical bases on the lower surface. To comply with the Continuum Theory for smectic liquid crystals, the layers have to be continuous throughout the preparation.

In the foregoing section on textures, the discussion has been based on a concept of rather finite smectic A layers. The reader must therefore be reminded of the comments made earlier in relation to Fig. 1.5 when the really diffuse nature of the layers was stressed, i.e. smectic A phases are in fact characterized by 1-d density waves with algebraic decay of positional correlations along the density wave.

Identification and classification of the smectic A phase

Microscopic textures

(a) The smectic A phase usually adopts one of two textures, namely the homeotropic or the focal-conic fan texture.
(b) The focal-conic fan texture always separates from the preceding phase or the isotropic liquid, on cooling, in the form of bâtonnets, which themselves consist of growing focal-conic domains. If the phase separates in streaks or droplets, it usually indicates that the material is impure. Full transition usually occurs over a fall in temperature of 0.3°C.
(c) The backs of the fans are usually (but not always) grained, due to the occurrence of parabolic defects (see Chapter 10). These grains give a radiating pattern along the length of a fan-shaped area; compare Plates 1 and 8. This graining, if not formed initially during the cooling cycle, can often be obtained, if the material exhibits a smectic B phase, by cooling the smectic A phase to the smectic B phase and then reheating. This then gives a good indication of both the smectic A and smectic B phase types and of the transition from smectic A to smectic B, particularly if no transition bars occur at the phase change.
(d) If the smectic A phase is first formed as a precursor for studies of a sequence of other phases, then to facilitate the identification of tilted phases, the smectic A phase is best obtained in the homeotropic texture; for the study of subsequent orthogonal phases, it is best to obtain the smectic A phase in its focal-conic fan texture.

Miscibility studies

(a) Mixtures of two materials each giving a smectic A phase may sometimes give rise to an increase in the thermal stability of the phase over a range of binary mixtures (see Fig. 1.20). Mixtures very rarely

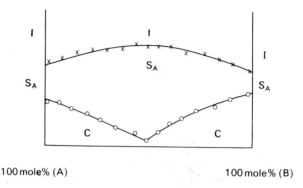

Figure 1.20 Miscibility diagram for two compounds A and B showing total mixing but enhanced S_A properties.

produce minima in the S_A-isotropic liquid transition lines that can make classification of an A phase difficult. Sometimes, a compound can give a very short temperature range A phase followed by a S_B phase; the transitions are conveniently classified as isotropic liquid-S_{AB} transitions and these are discussed in a later section. Co-miscibility can however usually be established for the smectic A phases of two different materials.

(b) Standard materials that exhibit the smectic A phase and are useful in miscibility studies are:

$$C_4H_9--N=CH--CH=N--C_4H_9$$

(i) *N*-Terephthalylidene-bis-4-n-butylaniline (TBBA) (for high temperature smectic A phases).

$$I \rightarrow N \rightarrow S_A \rightarrow S_C \rightarrow S_G \rightarrow S_H$$

(ii)
$$C_8H_{17}O--CN$$

4-Cyano-4'-n-octyloxybiphenyl (8OCB) (for compounds with terminal cyano-groups and interdigitated bilayer smectic A phases).

$$I \rightarrow N \rightarrow S_A$$

(iii)
$$C_2H_5O.OC--\underset{\downarrow}{N=N}--CO.OC_2H_5$$
$$O$$

Ethyl 4-azoxybenzoate (for smectic A phases which occur at moderate temperatures).

$$I \rightarrow S_A$$

X-ray diffraction pattern

A typical diffraction pattern for an unoriented smectic A phase is shown in Fig. 1.21. The outer ring is of a diffuse nature, thus characterizing the liquid-like arrangement of molecules in the layers. The sharp inner ring gives the lamellar spacing.

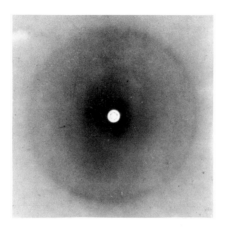

Figure 1.21 A typical X-ray diffraction pattern for a smectic A phase.

Enthalpy data from DTA or DSC

Differential thermal analysis (DTA) or differential scanning calorimetry (DSC) show that the smectic A phase gives quite a large enthalpy of transition to the nematic phase or the isotropic liquid. $\triangle H$ is usually about 1–1.5 kcal mol^{-1} (4–6 kJ mol^{-1}), indicating that these transitions are normally first order in nature (but there are a few cases where the transition can be much weaker and so considered second-order in nature).

References

Bonsor, D.H., Leadbetter, A.J., and Temme, F.P. (1978). *Mol. Phys.* **36**, 1805.
Bouligand, Y. (1972). *J. Phys. (Paris)* **33**, 525.
Cladis, P.E. (1975). *Phys. Rev. Lett.* **35**, 48.
Cladis, P.E., Bogardue, R.K., and Aadsen, D. (1978). *Phys. Rev. A.* **18(5)**, 2292.
De Vries, A., Ekachai, A., and Spielberg, N. (1979). *Mol. Cryst. Liq. Cryst.* **49**, 143.
Dianoux, A.J., Heidemann, A., Volino, F., and Hervet, H. (1976). *Mol. Phys.* **32**, 1521.
Friedel, G., and Grandjean, F. (1910). *Bull. Soc. franç. Minéral.* **33**, 192 and 409.
Friedel, G. (1922). *Ann. Physique* **18**, 273.
Grandjean, F. (1917). *Compt. rend. Acad. Sci. (Paris)* **166**, 165.
Gray, G.W., Harrison, K.J., and Nash, J.A. (1973). *Electron. Lett.* **9**, 130.
Hartshorne, N.H., and Stuart, A. (1970). 'Liquid Crystals'. In *Crystals and the Polarising Microscope*, 4th edn., Arnold, London. Chap. 14, pp. 509–523.
Leadbetter, A.J., Temme, F.P., Heidemann, A., and Howells, W.S. (1975a). *Chem. Phys. Lett.* **34**, 363.

Leadbetter, A.J., Richardson, R.M., and Colling, C.N. (1975b). *J. Phys. (Paris)* **36**, 37.
Leadbetter, A.J., Richardson, R.M., Dasannacharya, B.A., and Howells, W.S. (1976). *Chem. Phys. Lett.* **39**, 501.
Leadbetter, A.J., Durrant, J.L.A., and Rugman, M. (1977). *Mol. Cryst. Liq. Cryst. Lett.* **34**, 231.
Leadbetter, A.J., and Richardson, R.M. (1978). *Mol. Phys.* **35**, 1191.
Leadbetter, A.J., Frost, J.C., Gaughan, J.P., Gray, G.W., and Mosley, A. (1979a). *J. Phys. (Paris)* **40**, 375.
Leadbetter, A.J., and Richardson, R.M. (1979b). 'Incoherent quasielastic neutron scattering.' In G.R. Luckhurst and G.W. Gray (eds.), *The Molecular Physics of Liquid Crystals.* Academic Press, London and New York. Chap. 20, pp. 451–483.
Leadbetter, A.J. (1979c). *'Structural studies of nematic, smectic A, and smectic C phases.'* In G.R. Luckhurst and G.W. Gray (eds.), *The Molecular Physics of Liquid Crystals.* Academic Press, London and New York. Chap. 13, pp. 285–316.
Leadbetter, A.J. (1980). Private communication.
Lösche, A., Grande, S., and Eider, K. (1973). *First Specialised Colloque Ampère,* Krakow, Poland, p. 103.
Lösche, A., and Grande S. (1974). *18th Ampère Congress,* Nottingham, England, p. 201.
Madhusudana, N.V., Sadashiva, B.K., and Moodithaya, K.P.L. (1979). *Current Sci.* **48(14)**, 613.
Sackmann, H., and Demus, D. (1966). *Mol. Cryst. Liq. Cryst.* **2**, 81.

2 The smectic B phase

Introduction

Compounds exhibiting phases now known to be smectic B in type were first discovered as early as 1939, but, as with the smectic A phase, this phase was not assigned a separate code letter until 1966. The smectic B phase was then one of the three phases assigned a code letter in the original classification of smectic phases made by Sackmann and Demus (1966). Since this time, the smectic B phase has been the subject of many studies which have resulted in conflicting views about its true nature (crystal B phase or liquid crystal B phase).

Firstly, if we examine the history of the classification of the smectic B phase, we find that originally there were two types of phase called B. One had its constituent molecules arranged within the layers in a hexagonally close-packed array with the molecular long axes *perpendicular* to the layer planes; the other had a similar hexagonal packing arrangement, but the molecular long axes were *tilted* with respect to the layer planes.

Both of these phases were thought, and indeed assumed, to be co-miscible. Hence, in a number of early reports of new materials exhibiting either of these types of phase, both kinds were identified as smectic B phases. However, more recent studies by Doucet and Levelut (1977) and de Jeu and de Poorter (1977) using X-ray diffraction techniques, and by Goodby and Gray (1979) using miscibility methods, have conclusively proved that these two types of phase have separate identities and therefore should have separate code letters. Thus, the phase that is uniaxial has retained the designation (smectic B) originally made by Sackmann and Demus. The biaxial, tilted phase was then given the code letter H by many workers, because de Vries and Fishel (1972) had reported this phase in N-(4-n-butyloxybenzylidene)-4-ethylaniline (4O.2). Quite recently however, it has emerged that the relevant smectic phase of 4O.2 had been shown to be miscible with the smectic G phase first discovered and classified a year earlier by Demus and co-workers (1971) in the compound of structure

$C_5H_{11}O-\phenyl-C(=N)-N=\phenyl-C_5H_{11}$

On grounds of historical priority therefore, the smectic phase formerly called a tilted B phase should be called smectic G and *not* smectic H (Demus and co-workers, 1980). Conversely, phases classed by those outwith the group of Sackmann and Demus at Halle as smectic G should now be classed as smectic H. These points are discussed more fully in later sections.

Thus, a large number of early (1966–1972) reports of the classification of smectic phases as B may conflict with the current system of nomenclature. The situation was alleviated somewhat in the early 1970s by members of the Orsay research group who designated the uniaxial phase (smectic B) as S_{B_A} and the biaxial phase (smectic G) as S_{B_C}, i.e. they used the subscripts A and C by analogy with the orthogonal A and tilted C phases (see later). However, the two phases S_{B_A} and S_{B_C} have only really been referred to as having separate identities since 1978–1979, and as stated above, only since 1980 has general agreement been reached on their classification as smectic B and smectic G, respectively.

Secondly, there have been numerous problems concerning the identification of both the B (B_A) and the G (B_C) phases by miscibility methods. Original studies had always shown the two phases to be co-miscible; however, more detailed investigations have shown that the two phases are indeed immiscible. Furthermore, the immiscible region is often very small (usually of the order of 2–3% of the composition range of the two-component mixture) and is sometimes missed in complex cases or when contact preparations are used. It has also been shown in some miscibility studies involving two smectic G compounds that certain binary mixtures exhibit smectic B phases. Hence, two tilted G (B_C) phases can produce a smectic B (B_A—uniaxial) phase under suitable conditions of admixture. The smectic B phase has also been shown to have some affinity for the smectic F and I phases under certain conditions of mixing, and these phases too can appear to be miscible.

Thirdly, the study of the structure of the B phase has become more detailed with time, and therefore progressively varying models for the phase have been produced. Originally, the phase was shown to have its constituent molecules packed in a hexagonal array with their molecular long axes perpendicular to the layer planes. The molecules were assumed to rotate and the layers to be free to slide over one another, i.e. the system was considered as a *true* liquid crystal. However, recent studies have shown in *some* cases that not only is the hexagonal close-packed arrangement within a given layer very long-range, but also the hexagonal net is correlated over an extremely large number of layers. Thus, in some B modifications, the *undisturbed* phase is structurally solid-like and not simply a liquid crystal phase—see also Chapter 10, p. 134.

However, some of the physical properties of these crystal B phases are more consistent with those of a true liquid crystal modification than with those of a solid. For example, the tilted version of the phase for mixtures can be chiral in character when the constituent molecules are optically active (i.e. are of a chiral nature); the phase also exhibits shear and flow properties under stress; the transitions to and from the phase (i.e. at the upper temperature limit of the phase) are *always precisely reversible,* and do not undergo supercooling; in numerous homologous series of compounds that exhibit smectic B phases, the transition temperatures marking the upper limits of the phases show *regular trends* with definite *alternations*; finally, the Mössbauer spectra of various smectic B phases are inconsistent with the phase being a solid. These characteristics are *not* found when we are dealing with solids, but all are normally observed when we are dealing with true liquid crystals.

Despite this, current opinion was beginning to harden to the view that all B phases were of this very ordered (3D) type, until a report by Leadbetter, Frost, and Mazid (1979*a*) which showed that some B phases involve inter-layer correlations over, at the most, one or two layers, or possibly have no inter-layer correlations at all. Such B phases must be regarded as true liquid crystals.

It is against this somewhat confusing background that we will try to explain the nature, structure, and texture of the smectic B phase in the following sections.

Structure of the smectic B phase

Early X-ray diffraction studies on unaligned (powder) samples, coupled with conoscopic microscopic investigations of the smectic B phase, showed that the molecules were arranged in layers with the molecular centres positioned in a hexagonally close-packed array. The molecular long axes were shown to be orthogonal to the layer planes, and an indication of this can be drawn from the fact that the phase exhibits a homeotropic texture and is positive uniaxial. The molecules were also shown to be rotating about their molecular long axes quite rapidly. A simplified picture of the structure of the phase is shown in Fig. 2.1.

The first classical X-ray diffraction study of the smectic B phase was carried out by Levelut and Lambert (1971). They chose to make a single crystal study of the orthogonal smectic B phase of EBAC (ethyl 4-(4′-ethoxybenzylideneamino)cinnamate) and of the tilted smectic B (S_{B_C} or smectic G phase) of TBBA (terephthalylidene-bis-4-n-butylaniline).

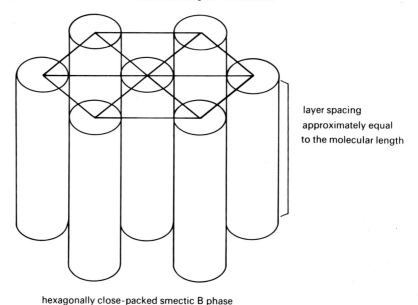

hexagonally close-packed smectic B phase

Figure 2.1 The molecules are orthogonal to the layers and packed in a hexagonal array. The rotating molecules are depicted as smooth cylinders.

At this point we should re-emphasize that the tilted B, i.e. the S_G phase is distinct from the orthogonal B phase and has a separate identity; this phase will therefore be discussed in a separate section under the smectic G phase.

By mounting perfect single crystals (solids) of these materials in an X-ray diffractometer and heating to the required temperature, Levelut and Lambert were able to measure the unit cell dimensions for the smectic B phase. Their results proved to be extraordinary when related to the molecular dimensions of the compound under examination.

Firstly, for EBAC (the smectic B material), they were able to show that the lamellar spacing in the B phase corresponded to the length of the molecule. Thus, they confirmed that the phase was indeed uniaxial and that no interdigitation or tilting within the layers occurs as for the smectic A phase.

Secondly, they were able to show that the dimensions of the hexagonally close-packed net within the layers were of a size such that *free* rotation of the molecules about their molecular long axes was impossible. This second observation poses a number of problems when we consider the rotational disorder of the molecules within each layer.

If we assume the molecules in the B phase have a blade-like, oblong structure, then when they rotate about their molecular long axes, the volume of space swept out is cylindrical. Thus, we could pack these cylinders closely together as shown (see upper part of Fig. 2.2). This represents the case when the edge of one blade-like molecule can just pass the edge of a neighbouring molecule, i.e. an edge-on-edge-on approach. This will be the limiting case for each molecule to have independent rotational disorder, such that its rotation will in no way be related to that of its nearest neighbour.

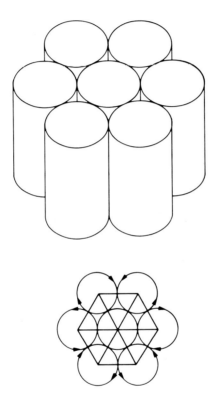

Figure 2.2 The close-packed hexagonal array of molecules showing the problems associated with free rotation of the molecules. The dimensions of the hexagonal net observed are smaller than those required in the lower part of the figure.

It can be seen from the plan diagram of the cylinders in the lower part of Fig. 2.2 that the molecules at the corners of the hexagon can rotate either freely, if the space between the molecular centres is large enough (which it is not), or in a matching sense if the cylinders are just in contact. However, if this is the case, the cylinder in the centre has problems. It cannot rotate in any direction freely because the hexagonal net is too small, and its sense of rotation cannot match that of all of its neighbours. This is important, because it must be remembered that the molecules at the corners of one hexagon are the centres of neighbouring hexagonal areas.

It is obvious therefore that rotation cannot occur in this way and that it must be completely co-operative (synchronized or co-ordinated). In practice, rotation occurs completely co-operatively, except for a small number of molecules with enough excess of energy to rotate almost freely (Luz, Hewitt, and Meiboom, 1974).

Levelut and Lambert were able to show that the hexagonal net dimension for EBAC was 4.8Å (see Fig. 2.3). This dimension only just

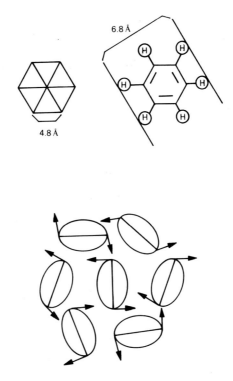

Figure 2.3 The net dimension for EBAC=4.8Å. This requires a co-ordinated rotation.

permits the molecules to rotate co-operatively such that the *edge* of one molecule is presented to the *side* of a neighbouring molecule (see Fig. 2.4, and lower half of Fig. 2.3). Therefore, the model of the smectic B phase stemming from Levelut and Lambert's work was one in which the molecules were orthogonal to the layer planes, with their molecular centres hexagonally close-packed. The molecules were shown to be rotating quite rapidly in this phase. However, the hexagonal net dimension is of such a

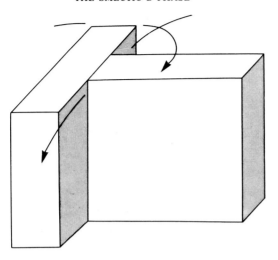

Figure 2.4 Edge-on-side-on approach of two rotating blade-like molecules in the smectic B phase.

size that the molecules must share space, and the rotation has to be of a co-operative nature. A further detail of their work was their interpretation of the diffuse scattering obtained in their X-ray diffraction experiments in terms of motions occurring between the layers. Their explanation for this was that columns of molecules (7–10 molecules long) stacked end-to-end were moving up and down between the layers as shown in Fig. 2.5, but this was shown later to be due to layer modulations.

With respect to the rotational situation, numerous workers have shown that co-operative rotation is theoretically possible in the smectic B phase. Seemingly, the description of this motion that emerges is one in which the rotation is restricted every 60° as the molecules pass each other in an edge-on-side-on approach. Thus, the motion is not smooth, but tends to slow very slightly as the molecules pass each other.

The current view of *some* smectic B phases is one that suggests that the phase is much more solid-like than was previously thought. Detailed X-ray diffraction studies on carefully aligned bulk samples of N-(4'-n-alkoxy-benzylidene)-4-n-alkylanilines (nO.ms) by Leadbetter and co-workers ((1979a, b) and (1980)), and on thin films by Moncton and Pindak (1979) have shown that the smectic B phases of these materials have three-dimensional order. Both sets of workers confirmed that the smectic B phase has its constituent molecules arranged in a hexagonally close-packed array within the layers, and that the molecular long axes are perpendicular to the layer planes. They also showed that there is long-range hexagonal order within the layers. Thus, each molecule has its own position within the

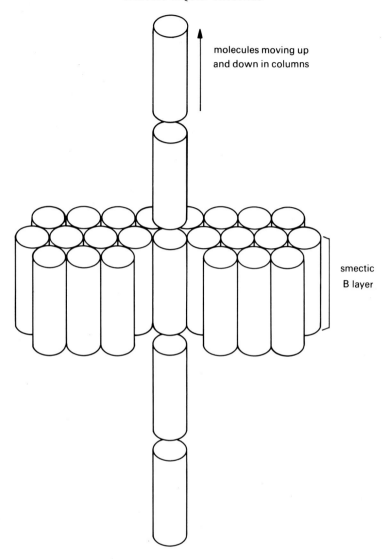

Figure 2.5 Levelut and Lambert's model for the B phase.

layer, i.e. a solid-like property. They were also able to show that the layers are highly correlated—another solid-like property. This structure is shown in Fig. 2.6.

This type of structure gives rise to a number of possible variations in the packing arrangements between the correlated layers. Leadbetter and co-workers (1979a, b) have shown that for a number of smectic B phases exhibited by the nO.ms (N-(4'-n-alkoxybenzylidene)-4-n-alkylanilines) the

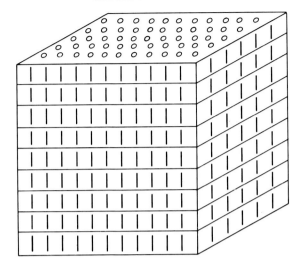

Figure 2.6 The correlated structure of the B phase—'infinite' order extending in three directions.

layers are arranged in such a way that for any particular layer, the molecules in the layer above and the molecules in the layer below lie over the same interstitial holes in the central layer (see Fig. 2.7). This type of bilayer packing arrangement is of the ABAB--- type, referred to as a cubic close packing of spheres in inorganic chemistry.

Furthermore, Leadbetter, Frost, and Mazid (1979a) showed that in certain cinnamate esters, an ABC--- trilayer packing arrangement between the layers could be obtained (see Fig. 2.8). With reference to a particular layer, the layer that lies above is turned through 60°, with respect to the layer lying below that particular layer, so that the upper layer covers the interstitial holes *not* covered by the lower layer.

An AAA---- monolayer packing is also possible, and there are also other mixed forms of these packing arrangements, e.g. an ABC---- packing of a random nature, i.e. ABCBACB etc.

In the smectic B phases of the nO.ms, all three packing types, AAA----, ABAB----, and ABCA--- have been found, and it is interesting to note that more than one stacking arrangement may occur at different temperatures for the smectic B phase of a given nO.m. For example, this change in stacking arrangement (Goodby and co-workers, 1980; Leadbetter, Frost, and Mazid, 1979a; Leadbetter and co-workers, 1979b) is exemplified by 5O.7 which changes from

bilayer → trilayer → bilayer
ABAB---   ABCA---   ABAB---

with decreasing temperature within the three-dimensional smectic B phase. No enthalpy changes or visible textural changes accompany these changes in stacking type (Gray, 1981).

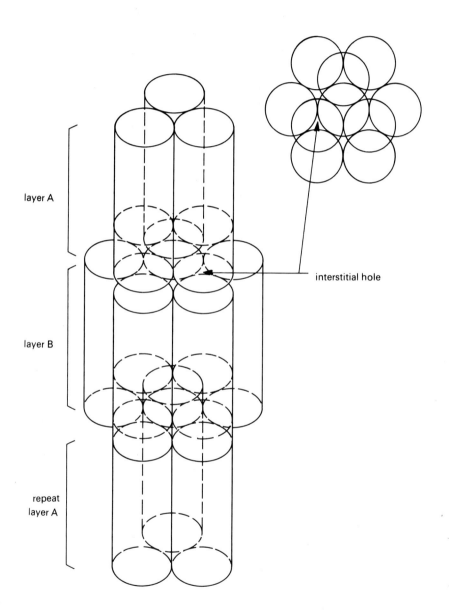

Figure 2.7 ABAB---- packing in the smectic B phase.

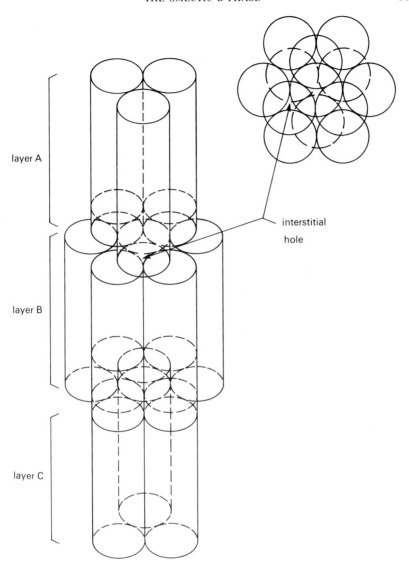

Figure 2.8 ABC---- packing in the smectic B phase.

A further important consequence of the work of Leadbetter and his colleagues was that if a smectic B phase forms a smectic G phase on cooling, then the B phase develops a dynamic wave motion within the layers. For example, for the nO.m, 5O.7, at a temperature about 7° above the S_B–S_G transition, the disc of diffuse scattering develops satellite peaks. These indicate a transverse modulation of the structure parallel to the layers having a wavelength about 17×the (100) spacing of the hexagonal net (see Fig. 2.9). At T_{B-G}, new peaks appear at a slight displacement from the satellites, and the central peak vanishes. Although the transition appears to be first order, the satellites are still given for 1 or 2° into the G (formerly known as the tilted B) phase. This suggests that the two phases can co-exist over about 3° around the transition. The wavelength of the undulatory motion is dependent on the material under examination, but is always of the order of 80Å.

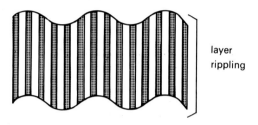

Figure 2.9 Undulatory motion in the B phase near the B–G transition or B–F transition.

These results demonstrate that throughout the B and G phase ranges, the direction of orientation of the molecular long axes is unchanged. At high temperatures, there are pronounced fluctuations which involve molecular displacements along the long molecular axial direction. As the temperature falls, these become less periodic, and the amplitude increases. The B–G change takes place when the molecules in adjacent (100) planes are displaced by about 2Å relative to each other. At the same time, the bilayer stacking arrangement of the B phase disappears. Since only one satellite appears, the modulations would appear to involve sinusoidal undulations of the layers (see Fig. 2.9). Presumably, these undulations trigger off the transition, and once the temperature has fallen about 3° below the B–G transition, the undulations cease and the layers become flat.

An interesting point concerning the B to G transition is, therefore, that the layers that form the G phase are *not* the same as the ones that existed in the B phase. The molecules do *not* tilt at the transition; the layers effectively tilt. This can be shown quite clearly by probe techniques (Luckhurst, 1979), which do not detect any tilting at the transition, and by the X-ray experiments which show changes in the diffraction pattern as the layers break and reform (see Fig. 2.10).

When the smectic B phase is *not* succeeded by a smectic G or F phase on cooling, then the B phase does not exhibit this undulating motion.

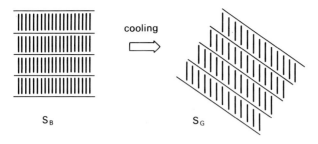

Figure 2.10 Smectic B to smectic G transition—the layers tilt, not the molecules.

Finally, we should emphasize that although the true nature—solid crystal or liquid crystal—of the B phase was at one stage open to question, the phase is now known to be capable of existing in either three-dimensional (crystal) or two-dimensional hexatic (liquid crystal) forms. The reader is here referred to Chapter 10 for recent developments.

Several of the structural details that have been discussed above can be related to features of the macroscopic textures of the B phase, and we will now try to illustrate this in the following section.

Textures of the smectic B phase

The natural textures of the B phase

There are two naturally occurring textures of the smectic B phase, namely the homeotropic texture and the mosaic texture. By the term 'natural texture', we mean those textures exhibited by the smectic B phase when it is formed on cooling either the isotropic liquid or the nematic phase. If the smectic B phase is obtained from the nematic phase, then the natural texture exhibited is dependent, to some degree, on the texture of the nematic phase. If the nematic phase is homeotropic, then the B phase will also exhibit a homeotropic texture; if however, the nematic phase is in its homogeneous or *schlieren* texture, then the B phase formed on cooling will usually exhibit the mosaic texture. One could therefore argue that these textures are paramorphotic textures predetermined by the nematic phase, but as both textures are *also* readily obtained by cooling the isotropic liquid, then it seems reasonable to define them as natural or spontaneous.

The homeotropic texture is of a similar nature to that formed by the smectic A phase. The layer planes of the molecules are parallel to the glass supports, and the texture is then viewed through a microscope perpendicularly to the layer planes—see Fig. 2.11. When viewed with the sample between two sheets of polarizer, which have their vibration directions at right angles to each other, the field of view is black. Conoscopic observation of this texture then reveals a positive uniaxial interference figure. An indication that the molecules are rotating rapidly in this phase can be drawn from the fact that it does exhibit uniaxial properties; if the molecules were merely oscillating, as in the smectic E phase (see later), the texture would then be birefringent, because the phase would have biaxial characteristics.

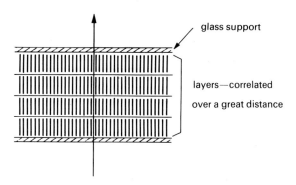

Figure 2.11 The homeotropic texture of the B phase. The layers may be correlated over large distances.

The mosaic texture of the B phase is the other natural texture. If the B phase is obtained directly from the isotropic liquid, the sample usually exhibits a mixture of the two natural textures, as shown in Plate 4 for 4-n-hexyl-4'-n-hexyloxybiphenyl. This is because the phase nucleates differently in separate areas. In this case, the majority of the sample has nucleated with the molecular long axes perpendicular to the glass surfaces, whilst in small areas (birefringent) some nucleation has taken place with the molecules at an angle to the surface. Some of the mosaic areas have a typical 'H' shape; these platelet areas are very common to the smectic B phase.

The mosaic texture obtained on cooling a nematic phase is shown in Plate 5 for 4-n-pentyloxybenzylidene-4'-aminobiphenyl. The mosaic area grows across the field of view as the nematic areas retract. The mosaic texture finally obtained for this material is shown in Plate 6, the mosaic platelets being of an oblong shape. We can also obtain mosaic textures in which the platelets are of a more rounded nature, as shown in Plate 7 for n-propyl 4-(4'-propylmercaptobenzylideneamino)cinnamate.

The structure of the mosaic platelets formed by the smectic B phase have not been investigated fully up to the present point in time. However, there would seem to be two possibilities for their structure; these are shown in Fig. 2.12(a) and (b).

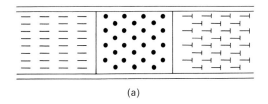

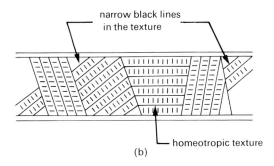

Figure 2.12 Possible arrangements for the mosaic texture of the B phase.

In both cases, except in any homeotropic area (Fig. 2.12(b)), the layer planes are *not* parallel to the glass supports, but are at some angle to them, such that the texture observed is birefringent. In the mosaic structure shown in Fig. 2.12(a), the constituent molecules have their long axes parallel to the supports, such that each mosaic area has some preferred direction of orientation of these axes, i.e. a preferred director.

In the second case, except in a homeotropic area, the layer planes or both the layer planes and the director corresponding to the molecular long axes are tilted at some arbitrary angle to the supporting surfaces.

Demus and Richter (1978) prefer the former explanation of the B mosaic structure, with the molecules *nearly* parallel to the supports; they also allow for some mosaic areas that are homeotropic.

If the wall separating two mosaic areas is normal to the supporting surface (Fig. 2.12(a)), the junction between the areas should appear as a black line. This is often, but not always observed. Sometimes, the junction is made up of several very narrow, dark lines. This would be explained either by the walls of Fig. 2.12(b), or by slightly tilted walls arising from model (a) in which the molecules are less than perfectly parallel to the supporting surfaces, cf. Demus and Richter (1978).

Since a large number of mosaic areas do show discontinuities made up of a number of narrow lines, these would be most easily explained by model (b), and since the differences between (a) and (b) are largely dependent on surface alignment, mosaics of either type may be possible, dependent on the compound and the boundary conditions. Also, if a smectic E phase underlies a smectic B phase on cooling, then the paramorphotic mosaic texture of the E phase is crossed with dark parallel lines indicating that contractions across the layers have taken place at the transition from the B phase to the E phase. Either model (a) or (b) would account for this effect.

The mosaic and homeotropic textures (involving *plane* layers) are, of course, more likely to be natural textures of the B phase than the focal-conic fan texture (involving *curved* layers), because the hexagonal, ordered structure of the B phase is more compatible with plane layered structures. The three-dimensional ordered structures of many smectic B phases will also be conducive to the formation of flat lamellar sheets in bulk samples rather than the *curved* arrangements of focal-conic fans. Clearly, it will be more difficult for the three-dimensional sheets (or blocks) to curve to form the fan texture, and so, for the B phase to exhibit the fan texture, it must be inherited from another phase on cooling, i.e. imposed upon the B phase—a condition referred to as a paramorphotic texture.

The paramorphotic focal-conic fan texture of the smectic B phase

As discussed above, the smectic B phase does not exhibit the focal-conic fan texture naturally; the texture is paramorphotic, and is only exhibited by the phase when it is inherited from the previous phase on cooling. For example, the B phase will exhibit the focal-conic fan texture when it is formed by cooling a smectic A or a smectic C phase which was already exhibiting that texture.

The structure of the focal-conic fan domain is the same as that for the smectic A phase (see Chapter 1), except that the layers may be correlated and the constituent molecules are hexagonally close-packed over extended distances within each layer. The microscopic appearance of the focal-conic domains is dependent on these two extra structural features. If we have correlation between the layers, and if this correlation is infinite, then we could assume that for the area of the fan at the glass boundary (the back of the fan), the layers will form a smooth surface. Thus, the appearance of the fan is that of a *smooth* cone with a very clear back, showing no blemishes or fissures. This type of fan texture is shown very clearly in Plate 8 for the crystal type B phase of n-decyl 4-(4'-phenylbenzylideneamino)cinnamate, and this can be compared with Plate 1 showing the same area for the smectic A phase. Note that there are no lines running from the apices of the fans in the B phase as there are in the A phase. This is again because the B phase has a more ordered structure and the backs of the fans are smooth. Plate 9 shows the fan texture of the stacked hexatic B phase of n-hexyl 4'-n-pentyloxybiphenyl-4-carboxylate (65OBC) p. 136. The fan texture of this more fluid version of the B phase is very similar to that of the more ordered B modification. These are probably the best features to observe when deciding whether a fan texture is of the A or B type.

There is another modification of the B focal-conic fan texture, and that is the 'stunted' or 'truncated' version. This texture is usually restricted to Schiff's bases, for example, the N-(4'-n-alkoxybenzylidene)-4-n-alkyl-anilines (nO.ms) and the 4-n-alkoxybenzylidene-4'-aminobiphenyls which exhibit crystal B phases.

In this texture the fans are slightly deformed and are not true conics. The edges of the fans become stepped and the apices are squared-off (truncated), and not pointed. Their appearance becomes somewhat more cubic in nature than conic. Plate 10 shows the truncated focal-conic fan texture for the smectic B phase of N-(4'-n-heptyloxybenzylidene)-4-ethyl-aniline (7O.2). This texture has only recently been noticed, and therefore, a full understanding of its structure has yet to be found. The 'stepped' nature of the edges of the fans may, however, be due to extensive, but not infinite, correlations between the layers. Thus one 'step' may correspond to one correlated stack of layers in the crystal type B phase—see pages 139 and 140.

Transition bars

These are lines which occur across the backs of the fans at the transition between a smectic A phase and a smectic B phase—see Plate 11 for n-decyl 4-(4'-phenylbenzylideneamino)cinnamate. These bars are of a completely transitory nature; at the point of transition the lines appear across the backs of the fans in concentric arcs, i.e. they run parallel to the layer planes, but as the transition reaches completion, the lines disappear and the backs of the fans heal up to produce the fan texture of the new phase (A or B).

The lines occur in a reversible way, but in some cases, after numerous reversals, they are replaced by parabolic focal-conic defects (wishbones) in the A phase; these defects then disappear and give the smooth fans of the B phase. This observation can also be made for some S_A–S_B transitions when the heating and cooling rates are very slow. The origins of these defects are discussed in a later chapter (p. 140).

There have been suggestions (Demus and Richter, 1978) that the occurrence of transition bars is dependent on the purity of the material under examination, i.e. the more impure a sample is, the more likely it is that one observes transition bars. However, using very pure compounds that exhibit S_A to S_B transitions, we have still observed transition bars. The transition bars can, however, be more pronounced in mixtures, as in Plate 12 for a binary system of terephthalylidene-bis-4-n-butylaniline (50 wt %) and n-hexyl 4'-n-dodecyloxybiphenyl-4-carboxylate (50 wt %).

The converse can also be true. For example, some of the n-alkyl 4'-n-alkoxybiphenyl-4-carboxylates that exhibit S_A to S_B transitions do not show transition bars at the point of transition, and in miscibility experiments involving these compounds with other standard S_A, S_B materials, the binary mixtures often still do not show transition bars.

Our current feeling is that transition bars are not due to impurities or other outside influences, but are due entirely to changes in structure and surface effects occurring at the point of transition—see also Chapter 10.

The isotropic to S_{AB} transition (I–S_{AB})

This type of transition (Coates and Gray, 1976) occurs when, on cooling the isotropic liquid, the B phase is formed via a *very* short-range smectic A phase. The B phase formed in this way exhibits unusual textures; the appearance of these textures is dependent on the temperature range of the A phase.

Firstly, an I–S_{AB} transition can be shown to exist by miscibility studies. If a miscibility study is carried out between a material that exhibits S_A *and* S_B phases, and one that exhibits an I–S_{AB} transition, then we would obtain a miscibility diagram of state, as shown in Fig. 2.13.

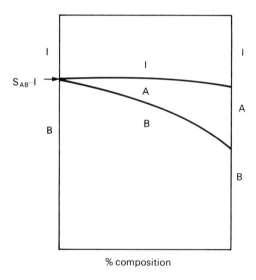

Figure 2.13 Miscibility diagram of state for an S_{AB}–I material.

In this diagram of state, the I–S_A and S_A–S_B transition temperature lines approach each other asymptotically to the point marked I–S_{AB} for the material on the left-hand axis, i.e. the S_A temperature range is very narrow.

The B phase that results from this type of transition can exhibit two textural forms depending on the narrowness of the temperature range of the smectic A phase, and on the correlations between the molecular layers within the phase.

If the A phase is extremely short-range, then the texture of the B phase that results has a square-shaped fan texture and a moss-like texture (Plate 13). The fan texture starts to form normally in the A phase, but as it develops, it goes through the transition to the B phase which prefers to adopt a mosaic texture. As a result, the fans set in mosaic shapes. The mossy areas result either from those parts of the A texture which have only just formed homeotropic areas, or by nucleation from the isotropic liquid directly to S_B. On transition to the B phase, very small mosaic areas form in the homeotropic or isotropic regions, thus giving a moss-like appearance. Plate 14 shows the separation of the smectic B phase from the isotropic liquid via an infinitely short smectic A phase for 3-methylbenzyl 4-(4'-phenylbenzylideneamino)cinnamate.

When the A phase is of a more finite temperature range than in the preceding case, then we obtain a different type of texture. This time the texture of the A phase has had time to set before transition to the B phase occurs. Hence the textures obtained for the B phase are similar to those normally exhibited by the phase (paramorphotic fan or homeotropic). There are slight differences, however, in the way in which the texture forms. The phase separates from the isotropic liquid in the form of bâtonnets and spherulites. These usually coalesce and tend to develop together to form rod-like growths. Thus, the fans form in rows and discs rather than individually. Plate 15 shows the separation of the B phase from the isotropic liquid, the transition occurring through a finite temperature range S_A phase for methyl 4'-n-octyloxybiphenyl-4-carboxylate.

The smectic B phase of trans,trans-4-n-propylbicyclohexyl-4'-carbonitrile

This material, code named CCH-3 (Pohl and co-workers, 1978), exhibits three smectic phases and a nematic phase. The nematic phase gives a smectic B phase on cooling. The B phase is unusual in that it normally separates from the nematic phase in only a small number of domains—see Plate 16. Thus, the mosaic platelets are of a very large size, see Plate 17. The nature of the phase is also unusual in that it is not miscible with conventional smectic B materials that we have used, and it forms monodomain samples quite easily. The nature of the phase has now been investigated fully, and the results (Brownsey and Leadbetter, 1981) of the X-ray diffraction studies reveal some unusual structural features not normally found in other smectic B phases. These indicate that the B phase has a bilayer structure—see p. 149.

Identification and classification of the smectic B phase

Microscopic textures*

(a) The smectic B phase usually exhibits two natural textures (the homeotropic and mosaic textures) and one paramorphotic texture (the focal-conic fan texture). The homeotropic B texture can be obtained as a paramorphotic texture as well as a natural texture. On standing or annealing for a long while, the paramorphotic fan texture can form the mosaic texture.

(b) The phase separates from the isotropic liquid in the form of platelet discs or oblong sheets. These areas will often be surrounded by a homeotropic region. Some of the birefringent mosaic areas will be 'H' shaped, a characteristic of the phase.

(c) In the focal-conic fan texture, the backs of the fans appear very clear and do not show any deformities, cracks, or blemishes. If the phase is formed on cooling a S_A phase, which itself was formed by cooling the isotropic liquid *slowly*, then the fans of the A phase will also have smooth surfaces, so that there will be little change on cooling to the B phase. However, on reheating to the A phase, the fans *will* show blemishes and lines radiating from the apices of the focal-conic domains.

(d) The phase is more viscous than the A phase, but still shows shear flow.

Miscibility studies

(a) Miscibility between two B phases produces very few problems and is usually straightforward. It is important to note that, if we are studying a material that does not show transition bars at a S_A to S_B transition, an appropriate standard material that shows either *good* transition bars or a wide temperature range S_C phase above the S_B phase is useful for miscibility studies.

(b) The B phase can often be injected into binary mixtures involving two smectic G (formerly called tilted B) phases. It can also be injected into mixtures involving materials having smectic F or I phases. These types of problem naturally complicate any miscibility study.

---

*Textural differences between crystal B and hexatic B phases are discussed fully in Chapter 10.

(c) Standard S_B materials that are useful in miscibility studies:

(i) $\text{n-C}_4\text{H}_9\text{O}-\bigcirc\!\!-\text{CH=N}-\bigcirc\!\!-\text{C}_8\text{H}_{17}\text{-n}$

N-(4'-n-butyloxybenzylidene)-4-n-octylaniline (4O.8) (standard three-dimensional, correlated B phase).
$$I \to S_A \to S_B$$

(ii) $\text{n-C}_6\text{H}_{13}\text{O}-\bigcirc\!\!-\bigcirc\!\!-\text{C}_6\text{H}_{13}\text{-n}$

4-n-Hexyl-4'-n-hexyloxybiphenyl (standard for the S_B–I transition).
$$I \to S_B \to S_E$$

(iii) $\text{n-C}_8\text{H}_{17}\text{O}-\bigcirc\!\!-\bigcirc\!\!-\text{CO.O}-\bigcirc\!\!-\text{OC}_6\text{H}_{13}\text{-n}$

4-n-Hexyloxyphenyl 4'-n-octyloxybiphenyl-4-carboxylate (wide S_C temperature range above a two-dimensional S_B phase).
$$I \to N \to S_A \to S_C \to S_B$$

(iv) $\text{n-C}_5\text{H}_{11}\text{O}-\bigcirc\!\!-\bigcirc\!\!-\text{CO.OC}_6\text{H}_{13}\text{-n}$

n-Hexyl 4'-n-pentyloxybiphenyl-4-carboxylate (65OBC) (two-dimensional hexatic S_B phase).
$$I \to S_A \to S_B \to S_E$$

X-ray diffraction pattern

A typical X-ray diffraction powder pattern for a smectic B phase is shown in Fig. 2.14. The outer ring is very sharp and the inner ring is well defined.

Enthalpy data

Transitions to and from the smectic B phase are usually characterized by large enthalpy peaks obtained by DSC or DTA. Transitions to the phase are usually first order in nature, and the enthalpy of transition is of the order 1 to 2 kcal mol^{-1} (4–8 kJ mol^{-1}). Transitions to the two-dimensional B phase may however be second order in nature.

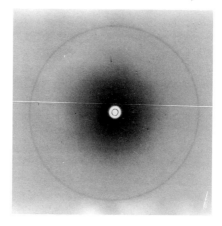

Figure 2.14 A typical X-ray diffraction pattern for an unoriented smectic B phase.

References

Brownsey, G.J., and Leadbetter, A.J. (1981). *J. Phys. (Paris) Lett.* **42**, 135.
Coates, D., and Gray, G.W. (1976). *The Microscope* **24**, 117.
De Jeu, W.H., and de Poorter, J.A. (1977). *Phys. Lett.* **61A**, 114.
Demus, D., Diele, S., Klapperstück, M., Link, V., and Zaschke, H. (1971). *Mol. Cryst. Liq. Cryst.* **15**, 161.
Demus, D., and Richter, L. (1978). 'The textures of smectic B.' In *Textures of Liquid Crystals*, V.E.B. Deutscher Verlag für Grundstoffindustrie, Leipzig. Chap. 4.7, pp. 86–88.
Demus, D., Goodby, J.W., Gray, G.W., and Sackmann, H. (1980). *Mol. Cryst. Liq. Cryst.* **56**, 311.
De Vries, A., and Fishel, D.L. (1972). *Mol. Cryst. Liq. Cryst.* **16**, 311.
Doucet, J., and Levelut, A.-M. (1977). *J. Phys. (Paris)* **38**, 1163.
Goodby, J.W., and Gray, G.W. (1979). *J. Phys. (Paris)* **40**, 363.
Goodby, J.W., Gray, G.W., Leadbetter, A.J., and Mazid, M.A. (1980). 'The smectic phases of the N-(4-n-alkoxybenzylidene)-4'-alkylanilines (nO.m's)—some problems of phase identification and structure.' In W. Helfrich and G. Heppke (eds.), *Liquid Crystals of One- and Two-Dimensional Order, Springer Series in Chemical Physics 11*, Springer-Verlag, Berlin, Heidelberg, and New York, pp. 3–18.
Gray, G.W. (1981). *Mol. Cryst. Liq. Cryst.* **63**, 3.
Leadbetter, A.J., Frost, J.C., and Mazid, M.A. (1979a). *J. Phys. (Paris)* **40**, 325.
Leadbetter, A.J., Mazid, M.A., Kelly, B.A., Goodby, J.W., and Gray, G.W. (1979b). *Phys. Rev. Lett.* **43**, 632.
Leadbetter, A.J., Mazid, M.A., and Richardson, R.M. (1980). 'Structures of the smectic B, F, and H phases of the N-(4-n-alkoxybenzylidene)-4'-n-alkylanilines and the transitions between them.' In S. Chandrasekhar (ed.), *Liquid Crystals*, Heyden, London, Philadelphia, and Rheine. pp. 65–79.
Levelut, A.-M., and Lambert, M. (1971). *Compt. rend. Acad. Sci. (Paris)* **272**, 1018.
Luckhurst, G.R. (1979). Unpublished results.
Luz, Z., Hewitt, R.C., and Meiboom, S. (1974). *J. Chem. Phys.* **61**, 1758.
Moncton, D.E., and Pindak, R. (1979). *Phys. Rev. Lett.* **43**, 701.
Pohl, L., Eidenschink, R., Krause, J., and Weber, W. (1978). *Phys. Lett.* **65A**, 169.
Sackmann, H., and Demus, D. (1966). *Mol. Cryst. Liq. Cryst.* **2**, 81.

3 The smectic C phase

Introduction

The smectic C phase, as with the A and B phases, was discovered a number of years before it was classified by a code-letter. The C phase was given this letter of identification in the original classification of smectic phases by Sackmann and Demus (1966). However, there are numerous reports in the earlier literature of materials that were later found to exhibit the smectic C phase, e.g. the 4-n-alkoxybenzoic acids (Bennett and Jones, 1939).

The smectic C phase was the first classified phase in which the molecular long axes of the constituent molecules were found to be tilted with respect to the normal to the layer planes. Furthermore, the molecules were found to be packed in an unstructured way within the layers; the S_C phase is therefore the tilted analogue of a smectic A phase. (It will be remembered of course that, though locally tilted, the average tilt angle in a smectic A is averaged to zero, whereas in a smectic C, the tilt direction is constant over considerable volume elements.) This led to many investigations of the nature of the tilting of the constituent molecules, and these studies gave rise to a number of theories about the mechanism and origin of the tilting process. At least seven separate theories (more than for any other smectic phase) have been derived, of which most depend on dipole-dipole interactions to produce the necessary tilting. However, all these theories have their drawbacks, and *no* one theory gives a complete explanation of the mechanism of tilting of the molecules.

In recent years, there have been various studies of the effects of changes in molecular structure on the incidence and temperature dependence of the smectic C phase; the results of this work have shown that the existence of the phase is also dependent on the molecular structure of the compound under investigation.

Although our understanding of the tilting of the molecules in the smectic C phase is by no means complete, all of the theories and scientific facts concerning the phase help to generate fresh ideas as to its true nature.

In the following text we will review the better-known theories of the C phase in the light of more recent experimental investigations.

Structure of the smectic C phase

Conoscopic microscopy shows that the C phase exhibits a biaxial interference figure, so indicating that the phase has its constituent molecules tilted at an angle, θ (defined as the tilt angle), to the layer normals. X-ray diffraction studies indicate that in this phase the molecules are arranged in layers, and that the molecular centres are packed in a random way. X-ray diffraction experiments also confirm that the molecular long axes are tilted with respect to the normal to the layer planes (lamellar spacing < molecular length; $d<l$). It is assumed that the tilt directions of the molecules in one homogeneous domain of a smectic C phase are aligned in the same direction, i.e. in the planes of individual layers and on passing in a direction at right angles to the layers through a succession of layers. Obviously, the tilt directions of the molecules cannot be randomly orientated, as this would result in an average tilt of zero, or homeotropy, i.e. a similar situation to that which arises in the smectic A phase.

It should be noted however, that in unaligned or poorly aligned samples in which the layer planes still lie parallel to the supporting surfaces, and correlation of the tilt direction persists from layer to layer, the tilt direction may change continuously over the area of the sample, except at line singularities or inversion walls where the tilt direction changes suddenly. This point arises later in discussions of the textures of smectic C phases. A simple structural diagram of the smectic C phase is shown in Fig. 3.1.

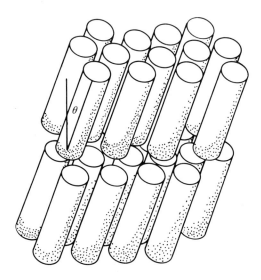

Figure 3.1 Structure of the smectic C phase; there is no regular arrangement of the molecular centres in the planes of the unstructured layers. Tilt angles (θ) may range upwards from $\lesssim 10°$.

THE SMECTIC C PHASE

The smectic C phase is of higher entropy than the B phase, but lower than the A phase. This means that by cooling it is possible to obtain the phase sequence $S_A \rightarrow S_C \rightarrow S_B$, i.e. orthogonal→tilted→orthogonal phase. The molecules in certain smectic A and smectic B phases have been shown to rotate rapidly about their molecular long axes, and therefore, it would be reasonable (page 53) that this is also the case in the C phase. Furthermore, because the molecules have an organized tilt direction, it would be unusual if any gyroscopic motion were involved.

Thus, the structural picture of the C phase is one in which the molecules are arranged in layers which are free to slide over one another, i.e. there is no long range correlation, except of tilt direction, between the layers. The constituent molecules are randomly packed within the layers, and the long axes of the molecules are tilted with respect to the layer normal. The molecules rotate about their long axes (with no associated precessional rotation), and thus the phase is biaxial. We are therefore dealing with a phase consisting of 2D unstructured layers with a tilted molecular arrangement.

The tilt angle in the C phase has been shown to vary with temperature, and it often increases with decreasing temperature in a uniform way (see Fig. 3.2). Similarly, for TBBA, θ changes gradually from $\sim 0°$ at the

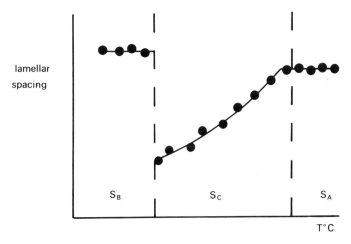

Figure 3.2 Temperature dependence of the tilt angle of the smectic C phase shown as a function of the change in lamellar spacing with temperature. Redrawn from Gray and Goodby, *Mol. Cryst. Liq. Cryst.* (1976).

S_A–S_C transition to $\sim 25°$ at the S_C–S_G transition (Doucet and co-workers, 1973). Some studies have suggested that there are two types of C phase, one in which the tilt angle is constant with respect to changing temperature, and one in which it varies with temperature. However, detailed investigations have shown that the tilt angle varies in all smectic C phases. For those in which the tilt angle appeared to be constant, a small change with temperature may in fact be detected. For example, for 4,4'-di-n-heptyloxy- and -octyloxy-azoxybenzenes, which have a S_C–N transition, the molecules in the cybotactic clusters in the nematic phase have a tilt angle of about 30°, and in the S_C phase the tilt angle increases from this value only very slightly with falling temperature (Leadbetter, 1979). In other cases, it was concluded, e.g. from esr studies by Luckhurst and Timimi (1979), that the S_C phases simply did not have long enough temperature ranges for the changes in tilt angle to be *easily* detected by measurements of lamellar spacings.

As mentioned above, in the case of smectic C phases which change to A phases at a higher temperature, the gradual changes in tilt angle with temperature give a tilt angle in the C phase (near to the S_C–S_A transition) which is close to the spread of tilt angles that individual molecules may adopt in the S_A phase, e.g. TBBA. This being the situation, in those cases in which only small changes in θ occur on passing through the underlying S_C phase with falling temperature, or when the S_C range is short (<10°), we therefore find cases in which the lamellar spacing changes vary little over the range of the S_A and S_C phases, e.g. n-hexyl 4'-n-decyloxy-biphenyl-4-carboxylate (Leadbetter, 1979).

There is no evidence for a strong first-order behaviour in which the tilt angle jumps suddenly to a finite value at the S_A to S_C transition.

Theories of the smectic C phase

So far we have given a simple description of the structure of the smectic C phase as it is currently understood. Over recent years, there have been a number of attempts to describe the structure of the phase and the mechanism for the tilting of the molecules as the phase is formed. Most of these theories are based on physical assumptions about the molecular structures of the compounds that exhibit smectic C phases. This type of approach usually ends in a mathematical description of the phase. The following sections will attempt to interpret a number of these theories in a pictorial way, so that they give a more visual understanding of the phase and the related hypotheses.

(a) McMillan's model of the smectic C phase

McMillan's theory of the smectic C phase (McMillan, 1973) was probably the first theory of the mesophase to be readily accepted, and it was quickly adopted as the standard structural model for the C phase. However, more recent investigations concerning materials that exhibit smectic C phases have cast doubts on the theory. In some cases, the theory has been disputed quite vigorously, and as a consequence it is now not as generally accepted as it was.

McMillan's model relied on the knowledge, at that time, that nearly all the materials that exhibited S_C phases had molecular structures that were approximately symmetrical in shape, e.g. the dimeric 4-n-alkoxybenzoic acids. He also noticed that *all* of the compounds had dipole moments associated with the *ends* of the central core structure, e.g. a large number of the compounds were of the n-alkoxy type. Thus, he assumed that for a compound to exhibit a smectic C phase it must have large terminal dipoles directed outwards from the molecular long axis and associated with electronegative atoms (e.g. oxygen, nitrogen, etc) situated at both ends of the central core structure. For example, the 4,4′-di-n-alkoxyazoxybenzenes (I) exhibit S_C phases, whereas the 4,4′-di-n-alkylazoxybenzenes (II) do not.

$$C_nH_{2n+1}O-\langle\bigcirc\rangle-N=N(O)-\langle\bigcirc\rangle-OC_nH_{2n+1} \quad\quad (I)$$

$$C_nH_{2n+1}-\langle\bigcirc\rangle-N=N(O)-\langle\bigcirc\rangle-C_nH_{2n+1} \quad\quad (II)$$

He defined the dipolar forces associated with these electronegative atoms as stemming from *terminal outboard dipole moments*. The lateral dipole moment associated with the central linkage (the azoxy linkage in (I) and (II)) he termed the *central dipole moment*.

In the resulting model, McMillan presents a physical picture of the smectic C phase in which the tilt angle plays only a secondary role. The primary role is played by a freezing out of the rotation of the molecules about their long axes in the nematic or smectic A phases at the transition to the C phase. The anti-parallel outboard dipole moments then become aligned, and a torque is created parallel to the layer planes, thus tilting the molecules with respect to the layer normal, as shown in Fig. 3.3.

By assuming that there are permanent dipole moments associated with the molecules (two outboard and one central dipole) and examining the rotational phase transition within the mean-field approximation, McMillan was able to show that his model would predict the formation of three more ordered smectic phases from one disordered smectic A phase (i.e. an A phase in which the dipoles are randomly organized within the plane of the layer). The first of the phases (C) predicted by this treatment has its central dipole much weaker than the two terminal dipoles, and thus has the physical properties associated with the smectic C phase (tilted director, optically biaxial, second order A to C phase transition). The second ordered phase (C_1) has its central dipole much stronger than the terminal

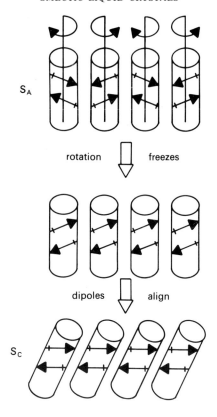

Figure 3.3 McMillan's model of the smectic C phase.

dipoles and therefore is an orthogonal ferroelectric phase. The third phase (C_2) has its (two terminal and one central) dipoles approximately equal in magnitude; thus the low-temperature phase is tilted and ferroelectric (see Fig. 3.4). In this way, the molecular rotation is assumed to freeze and the tilt angle passively follows the orientational order parameter.

Thus, it is from the internal balance of the permanent dipoles (two outboard and one central) associated with each molecule that a new magnitude of the dipole moments can be created and used to produce three smectic modifications. Note that the tilted smectic C phases developed in this way have the two outboard dipole moments pointing consistently in opposite directions, explaining the origin of the appropriate torque which has produced the tilting of the molecules. Although this mathematical treatment gives these three interesting models, it is unlikely that S_{C_1} in particular could ever exist. However, there are numerous compounds that exhibit S_C phases that can fall into the S_C and S_{C_2} categories of dipolar orientation.

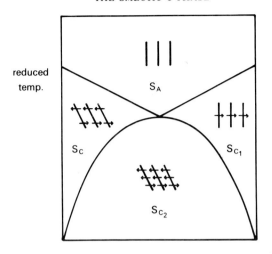

Figure 3.4 McMillan's model of the smectic C phase. The diagram shows the variety of possible S_C phases produced by the theory. Redrawn from McMillan, *Phys. Rev.* (1973).

(b) Wulf's model of the smectic C phase

Unlike McMillan's model, Wulf's theory is based purely on steric factors, i.e. on the shapes of the molecules (Wulf, 1975), and, as shown in Fig. 3.5, on the packing requirements needed for the zig-zag shaped molecules that often exhibit smectic C phases. The hypothesis also assumes that, besides having a zig-zag shape, the constituent molecules must be approximately symmetrical, a property at that time taken to be common to smectic C materials. The zig-zag shape of the constituent molecules may arise from obliquely directed end chains.

Wulf assumes that the interaction between such zig-zag molecules depends on mixed tensors $\vec{V}^3\vec{V}^2$, etc, as well as on the more usual tensors $\vec{V}^3\vec{V}^3$, $\vec{V}^2\vec{V}^2$ and other similar, higher order tensors, where $\vec{V}^3$ and $\vec{V}^2$ are unit vectors denoting the directions of the molecular long axis and a transverse axis, respectively.

He then finds a simple interaction, which includes a term that mimics the effect of the zig-zag shape of the molecules. By using values for the magnitude of the zig-zag term, he finds a second-order smectic A to smectic C transition.

Schematic representations of examples of the packing arrangements of the zig-zag molecules are shown in Fig. 3.5.

When these zig-zag shaped molecules pack together, they form layers in which the long axes are tilted with respect to the normal to the layer planes. Thus, the molecules form a smectic C phase.

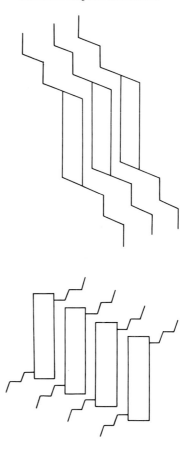

Figure 3.5 Wulf's packing model of hypothetical zig-zag shaped molecules in order to produce a smectic C phase. Redrawn from Wulf, *Phys. Rev.* (1975).

(c) Other theories of the smectic C phase

The Wulf and McMillan models express some of the diverse ideas concerning the reason why the molecules tilt in the smectic C mesophase. A number of later theories tend to extend and support either the dipole model or the steric model. For example, de Jeu (1977) showed that certain alkyl and/or alkoxy substituted azobenzenes may exhibit, besides the nematic phase, smectic A or smectic C phases (or both), depending on the end substituents. His results suggested that the repulsions between the zig-zag shaped molecules do not play an important role in the formation of the smectic C phase. However, the results obtained were in closer agreement with McMillan's dipole model of the C mesophase, provided that the molecules have a random head to tail arrangement. Thus de Jeu rationalized his results for the C phase in terms of a dipole model, in an extension of the treatment of McMillan. However, for the cases in which the constituent molecules have only one outboard dipole moment, he showed that there were two possibilities for a model with dipole interactions. De Jeu's suggestion was that this situation could possibly provide a model for the smectic F phase.

A later publication by Van der Meer and Vertogen (1979) on induced dipoles in the C phase disputed the dipole model of McMillan because this relied on a freezing out of the molecular rotation in order to create the C phase. Such a quenching of molecular motion does seem unlikely, because neutron scattering measurements show that molecules in the C phase are rotating with an almost random orientational distribution (Dianoux and co-workers, 1976), on a time scale of 10^{-11}s. On the other hand, the theory of Priest (1975, 1976) does not rely on such a freezing out of the rotations. However, this theory, based on general coupling between second rank tensors, does not discuss the relation between the coupling constants and molecular structure. Van der Meer and Vertogen attach importance to structural studies such as those by Goodby, Gray, and McDonnell (1977) and de Jeu (1977), and note that these do not support the steric models of Wulf (1975) and of other similar theories, e.g. by Cabib and Benguigui (1977). Although the structural work by Goodby, Gray, and McDonnell (1977) had shown that *terminal outboard* dipoles are not essential for a smectic C phase to exist, the results did show that such dipoles do in fact influence the *thermal stability* of the C phase, and Van der Meer and Vertogen (1979) considered that these studies, together with the structural studies of de Jeu, still reveal the importance of transverse dipoles in the molecules.

Van der Meer and Vertogen (1979) then proceeded to develop a model in which a permanent dipole in one molecule induces a dipole in a neighbouring molecule. The position of the permanent dipole is important in determining the contribution of the anisotropic part of the induction forces, and they conclude that an acentral dipole, particularly an acentral transverse dipole is very important in this respect. They suggest that there is an optimal location for this dipole in the molecule, and that it is the alignment of the induced and the permanent dipoles that creates the force for tilting (Fig. 3.6), the resistance to tilt being contributed to by a combination of Van der Waals' forces and hard core repulsions.

Thus, the molecules are allowed to rotate freely in this situation and the model is flexible enough to make allowances for changes in molecular structure. For small changes in molecular structure (e.g. extension of a terminal carbon chain), the theory merely allows a small change in the positions of the dipoles (induced and permanent), and it is the strength of

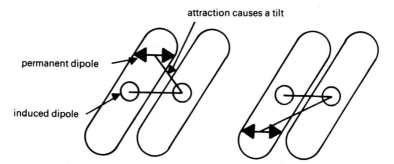

Figure 3.6 Model of Van der Meer and Vertogen for the smectic C phase. The tilt is caused by dipole-induced dipole interactions. Redrawn from Van der Meer and Vertogen, *J.Phys. (Paris)* (1979).

the dipole-dipole interaction that determines whether a smectic C phase is observed or not.

(d) Molecular structural factors influencing the occurrence of the C phase

There have been a large number of experimental studies concerning the effect of small changes in molecular structure on the incidence and temperature dependence of the smectic C phase. The conclusions that can be drawn from these studies are as follows:

(1) The smectic C phase is very dependent on the molecular structure of the constituent molecules, particularly on the length of the chains of the terminal alkyl or alkoxy groups.
(2) The phase appears to be exhibited predominantly by molecules that have two terminal alkyl chains (alkyl/alkoxy groups).
(3) The phase is usually exhibited by compounds that have molecular structures that are approximately symmetrical.
(4) Terminal chain branching can often increase the chances that a material will exhibit a smectic C phase.
(5) Terminal outboard dipole moments help to increase the chances that a compound will exhibit a smectic C phase; however, they are not prerequisites for formation of the phase.

The temperature dependence of the C phase on molecular structure can be demonstrated by examining the phase sequences exhibited by certain homologous series. For example, consider the n-alkyl 4'-n-decyloxy-biphenyl-4-carboxylates (Fig. 3.7) or the 4-n-alkoxyphenyl 4'-n-octyloxy-biphenyl-4-carboxylates (Fig. 3.8) studied by Gray and Goodby (1976*a*).

Smectic C phases are often suddenly injected at certain terminal alkyl chain lengths (e.g. at $n=4$ in Figs. 3.7 and 3.8). Indeed it appears that for esters of the general types shown in Figs. 3.7 and 3.8, when *one* terminal

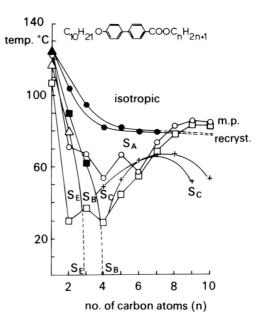

Figure 3.7
Plot of the transition temperatures against the number of carbon atoms (*n*) in the ester alkyl chain of the n-alkyl 4'-n-decyloxybiphenyl-4-carboxylates: ● S_A–I; ■ S_A–S_B–S_E; ▲ S_{AB}–Isotropic; + S_A to S_C or S_C virtual; □ mesophase-crystal on cooling; ○ crystal mesophase or isotropic on heating. Redrawn from Goodby and Gray, *Mol. Cryst. Liq. Cryst.* (1976).

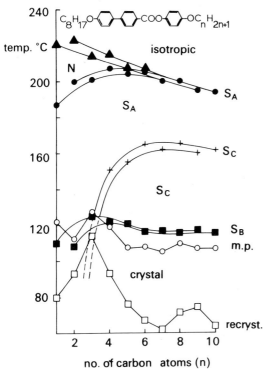

Figure 3.8
Plot of the transition temperatures against the number of carbon atoms (*n*) in the alkyl chain of the 4-n-alkoxyphenyl 4'-n-octyloxybiphenyl-4-carboxylates: ▲ N-isotropic; ● S_A to N or isotropic; + S_C to S_A; ■ S_B to S_A or S_C; ○ crystal to mesophase on heating; □ mesophase to crystal on cooling. Redrawn from Goodby and Gray, *Mol. Cryst. Liq. Cryst.* (1976).

chain reaches 7 or 8 carbon atoms in length, and the other terminal chain reaches 4 or 5 carbon atoms in length, the chances are that we will get an injection of smectic C properties. The situation is a little different with other materials. For example, with Schiff's bases, the C phase is often injected even earlier into homologous series.

Furthermore, with reference again to series of n-alkyl 4'-n-alkoxy-biphenyl-4-carboxylates, Goodby and Gray (1978) were able to show that the maximum point on the S_A–S_C transition temperature curve usually occurred at a value of $m=n-2$, where m is the number of carbon atoms in the terminal ester chain, and n is the number of carbon atoms in the terminal alkoxy chain. Thus, the highest-temperature smectic C phase is often exhibited by the member of the series which does not have the most symmetrical molecular shape. Indeed, it has a shape which is slightly 'unsymmetrical'.

With respect to the terminal dipole moments—the outboard dipoles of McMillan—a series of materials (III) to (VII) has been synthesized in which the terminal dipoles are progressively eliminated.

C_8H_{17}–⟨⟩–⟨⟩–CO.O–⟨⟩–OC_8H_{17} (III)

C_8H_{17}–⟨⟩–⟨⟩–CO.O–⟨⟩–C_8H_{17} (IV)

$C_2H_5\overset{*}{C}H(CH_3)(CH_2)_3$–⟨⟩–⟨⟩–CO.O–⟨⟩–$CH_2\overset{*}{C}H(CH_3)C_2H_5$ (V)

$C_2H_5\overset{*}{C}H(CH_3)(CH_2)_3$–⟨⟩–⟨⟩–CO.O–⟨⟩–$C_7H_{15}$ (VI)

C_8H_{17}–⟨⟩–⟨⟩–CO.O–⟨⟩–$CH_2CH(CH_3)C_2H_5$ (VII)

* chiral carbon atom
(VII) racemic modification

All the compounds exhibit smectic C or chiral smectic C phases. These results (Goodby, Gray, and McDonnell, 1977) would indicate that, although in some cases the occurrence (thermal stability) of the smectic C phase is aided by terminal dipole moments, they are by no means a necessity. In cases (V) to (VII) it would appear therefore that if McMillan's theory is correct, a sufficiently strong outboard dipole must mean one as small as the CH_3–C dipole of the branched alkyl chain. Even so, in (VII), only one such dipole is present. Alternatively, if Wulf's theory is correct, the shape of the branched chain systems must suit the steric requirements of the C phase.

Obviously, it is not possible to reach any definite conclusion at present based on the available structural information and the related theories. *No one piece of information or theory stands out above the rest as giving a well-defined explanation of the driving force that causes the molecules to tilt in the smectic C phase.*

Textures of the smectic C phase

The smectic C phase exhibits two microscopic textures, the *schlieren* texture and the focal-conic fan texture. The phase can exhibit both of these textures spontaneously when the phase is formed direct from the isotropic liquid or the nematic phase, but in the vast majority of cases, the *schlieren* texture is exhibited in preference to the focal-conic texture.

If the C phase is exhibited on cooling a smectic A phase, then the *schlieren* texture will be obtained from a homeotropic A texture, and the broken fan texture will be obtained from the clear, focal-conic texture of the preceding A phase. The fan texture can therefore occur naturally or paramorphotically.

The smectic C phase can be obtained only by cooling three other types of liquid crystal phase—the nematic, the smectic A, and the smectic D phases. Therefore, it can only inherit different textures from these few phases. Usually, in the case of precursor nematic and D phases, the C phase formed will exhibit the *schlieren* texture. The texture obtained from the A phase is, as stated above, either *schlieren* or focal-conic.

The schlieren *texture*

The *schlieren* textures of the nematic phase and the related C phase were investigated by Saupe (1973) and Nehring and Saupe (1972). Their treatment of the smectic C phase, although not as detailed as their classical treatment of the nematic phase, indicates nonetheless that the textures of the two phases are closely related.

A typical *schlieren* texture of the smectic C phase is shown in Plate 18 for 4-(2'-methylbutyl)phenyl 4'-(4"-methylhexyl)biphenyl-4-carboxylate. The black bands or *schlieren* occurring throughout the texture are regions of extinction and are often referred to as '*schlieren* brushes'—see Fig. 3.9. These brushes meet at point singularities on the surface of the preparation, or rather, point singularities are the origins of the brushes. The points are in effect the intersection of vertical lines of singularity with the surface.

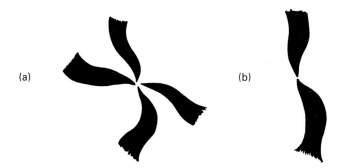

Figure 3.9 (a) point singularity with four brushes; (b) point singularity with two brushes.

For *schlieren* textures in nematics, two types of point singularity are observed, one in which two brushes originate from the centre, and one in which four brushes originate from the centre. For the smectic C phase, only centres with four derived brushes are observed. This is clearly shown in the *schlieren* texture of the smectic C phase of 4-n-hexyloxyphenyl 4′-n-octyloxybiphenyl-4-carboxylate (Plate 19). In both Plates 18 and 19, the birefringent areas between the *schlieren* or brushes are regions in which the tilt direction is at an angle to the vibration directions of the crossed polarizer and analyser.

Nehring and Saupe were able to show that the point singularities are characterised by

$$|s| = \frac{\text{number of brushes}}{4}$$

and have a positive sign when the brushes rotate in the same direction as that in which the polarizer and analyser are simultaneously rotated in the crossed position, and a negative sign when they turn in the opposite sense. Thus, singularities with

$$s = +\tfrac{1}{2}, -\tfrac{1}{2}, +1, -1$$

are known for nematics, but for the smectic C phase, all the point singularities are of the $s=\pm 1$ type. Singularities with $s=\pm\tfrac{1}{2}$ have never been observed, except for nematics. The sum of the positive and negative values of s averages out to a value of zero over a bulk sample.

From Nehring and Saupe's treatment there are a number of possible molecular arrangements associated with point singularities that can give rise to values of $s=\pm 1$.

Figure 3.10 illustrates the topological situation (the molecular alignment) near the centre of four point singularities such as may be present in a smectic C preparation, i.e. with $s=\pm 1$. The lines in Fig. 3.10 (and also Fig.

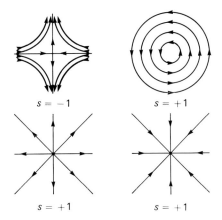

Figure 3.10 Examples of four possible topologies about point singularities found in the smectic C phase.

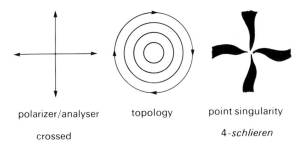

Figure 3.11 Examples of the origin of a point singularity in a smectic C phase.

3.11) represent the projections of the tilted molecules onto the x-y plane in which the layers of the smectic C phase lie. Thus, the tilt direction is changing continuously about the point singularity (a line singularity through the thickness of the film under observation). Since two tilt directions are possible in any one plane, it is necessary to represent the direction of tilt in some way. This is most simply done using an arrowhead to represent the uppermost ends of the molecules. Therefore, two situations are shown in Fig. 3.10 in which the tilt directions radiate from the centre—one in which the molecules lean away from the centre, and one in which they lean inwards, towards it. The other two distributions are self-explanatory, and all four would clearly give rise to four *schlieren* emanating from the point singularity, as represented in Fig. 3.11 for one of the topologies in Fig. 3.10.

Since lines or points with $s = \pm \frac{1}{2}$ are not compatible with the smectic C structure, the *schlieren* texture of the smectic C phase can be distinguished from that of the nematic phase by the fact that it exhibits only singularities that have four brushes originating from them.

The C phase can also exhibit two other variations of the *schlieren* texture. These are the sanded and the lined *schlieren* textures. The sanded texture tends to be exhibited by aromatic carboxylic acids and, for example, Plate 20 shows the *schlieren* texture obtained with 3'-nitro-4'-n-hexadecyloxybiphenyl-4-carboxylic acid. The point singularities and their related *schlieren* tend to be very small and blurred giving the impression of sanding.

The lined *schlieren* texture is often exhibited by materials that have long temperature range smectic C phases, and which are formed on cooling a smectic A phase. On cooling the C phase of such compounds, the birefringence colour of the texture changes as the tilt angle changes. Often at a certain temperature the domains between the *schlieren* become lined, as shown in Plate 21 for 4-n-hexyloxyphenyl 4'-n-octyloxybiphenyl-4-carboxylate. It has been suggested that these lines may be the result of the layers buckling under stress as the sample contracts on cooling, thus giving rise to a corrugation of the layers.

The focal-conic fan texture

The focal-conic fan texture *can* be exhibited by the smectic C phase, both naturally and paramorphotically, but it is very rare for this texture to be formed spontaneously.

The natural or spontaneous fan texture is not as broken as that exhibited paramorphotically. This would suggest that the tilt directions of the constituent molecules align themselves in a more uniform way when the fan nucleates from the nematic phase or the isotropic liquid.

The paramorphotic fan texture is always obtained on cooling the fan texture of a smectic A phase. The texture obtained in this way is broken and the fans have a sanded or grained appearance, as shown in Plate 22 for terephthalylidene-bis-4-n-butylaniline. Under high resolution and magnification, it is possible to observe by optical microscopy the effects of molecular fluctuations, i.e. the phase has a shimmering appearance. Furthermore, reference to the section on the fan texture of smectic A phases and attempts to fit a tilted arrangement of molecules into the focal-conic domains will clearly cause the reader to realize that packing problems must occur and lead to dislocation of the simple fan arrangement, i.e. to a breaking of the smooth fan texture.

The chiral smectic C phase

The smectic C phase exhibited by compounds that are of a chiral nature is itself optically active, i.e. the phase is termed a chiral smectic C. The structure of the optically active phase is essentially the same as that for the achiral C phase except for the distribution of the molecular tilt directions. If we consider a particular layer in a chiral smectic C phase, then it is assumed that the tilt directions of the molecules in the layers directly above and below are turned through small angles in opposite senses with respect to the tilt director of the reference layer—see Fig. 3.12.

In this way, the tilt directors form a helical distribution on moving from layer to layer. The helix has a pitch length that is temperature-sensitive because the pitch is itself dependent on the tilt angle of the molecules. As the helical pitch length changes with temperature therefore, so the optical properties of the smectic C phase vary. When the helical pitch length is of a suitable size, the phase will reflect iridescent light in the visible region of the spectrum, in a manner similar to a cholesteric liquid crystal. The wavelength and colour of the light reflected will then depend on the temperature of the phase. Thus, the chiral modification of a smectic C phase can be used as a temperature sensor in thermographic studies. In all the cases studied so far, the colour of light reflected from a chiral smectic C phase has changed with temperature in the opposite manner to that of the

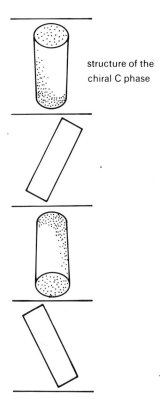

Figure 3.12 Helical structure of the smectic C phase (chiral).

cholesteric phase occurring at higher temperatures. In those cases, the light reflected from the cholesteric phase as the temperature falls has changed from blue to red, followed by a shift from red to blue on cooling through the range of the chiral smectic C phase. This is consistent with

(a) an increasing tilt angle and a decreasing pitch of the chiral smectic C as the temperatures fall, and
(b) the applicability of a $\lambda \propto P$ relationship.

The smectic C phases of chiral compounds show some unusual dipolar effects which stem directly from the chirality of the materials. Meyer and co-workers (1975) first demonstrated that a spontaneous, net polarization parallel to the layers occurs in chiral smectic C phases. They pointed out that achiral C phases have only a two-fold rotation axis parallel to the layers and normal to the molecular long axis, a reflection plane normal to the two-fold axis, and a centre of inversion. However, in the chiral phases, composed of chiral molecules, the mirror plane and the inversion centre are eliminated. The two-fold axis remains. When the molecule is non-chiral, the mean orientation of its permanent dipole must be normal to the two-fold axis, whereas the asymmetry of a chiral molecule forces the permanent dipole to have a component parallel to the two-fold axis. If all the molecules are identical, then a net polarization—a spontaneous dipole moment—parallel to the two-fold axis exists. Therefore, because of symmetry considerations, chiral C phases are ferroelectric. This behaviour was demonstrated by these authors (see also Moncton and co-workers, 1980) for

$$\text{n-}C_{10}H_{21}O-\!\!\left\langle\!\!\bigcirc\!\!\right\rangle\!\!-CH=N-\!\!\left\langle\!\!\bigcirc\!\!\right\rangle\!\!-CH=CH-CO.OCH_2\overset{*}{C}H(CH_3)C_2H_5$$

If the coupling of the molecule to the environment is weak and the molecule is almost freely rotating, Meyer and co-workers pointed out that the magnitude of the spontaneous polarization and the ferroelectric behaviour would be reduced. In an attempt to diminish this averaging effect, Keller and co-workers (1976) examined some other chiral C systems including

$$\text{n-}C_6H_{13}O-\!\!\left\langle\!\!\bigcirc\!\!\right\rangle\!\!-CH=N-\!\!\left\langle\!\!\bigcirc\!\!\right\rangle\!\!-CH=CH.CO.OCH_2\overset{*}{C}HCl.CH_3$$

and obtained encouraging results suggesting that increasing the dipole at the chiral centre did augment the ferroelectric behaviour.

Recent reports indicate that Clark and Lagerwall (1981) have produced a fast switching electro-optical device based on the influence of an external field on the spontaneous net polarization in the chiral smectic C phase. The existence of this net polarization parallel to the layers and the evidence from the work of Keller and co-workers implies that there must be a slowing down in the rotation of the molecules about their long axes at the point of dipole alignment; this could aid the tilting of the molecules. Gray and McDonnell (1976b) and Goodby, Gray, and McDonnell (1977) did however show that racemic systems—related to chiral systems exhibiting chiral C phases—form equivalent achiral C phases. Therefore, although the spontaneous polarization may *aid* tilting, it is not a primary factor in the formation of the phase. It plays a secondary role, i.e. it develops as a result of tilting.

The chiral smectic C phase, because of its unusual structure, exhibits a number of microscopic textures different from those of the achiral smectic C phase; these may be listed as follows.

(1) The pseudo-homeotropic texture

This texture is usually obtained on cooling the homeotropic texture of a preceding smectic A phase. Although the molecules tilt in the C phase (normally giving a birefringent achiral *schlieren* C texture), in the chiral C phase, the tilt directors are organized to give a helix. With the smectic layers parallel to the slide, the pitch direction of the helix is perpendicular to the glass supports, and therefore the tilt directions average out to zero; this gives a homeotropic type of texture. The texture is not a true homeotropic texture since it is being viewed only along an averaged optic axis, and it can co-exist quite naturally in contact preparations with the *schlieren* texture. Thus we have designated the texture as pseudo-homeotropic. The normal *schlieren* (C) and the pseudo-homeotropic (chiral C) textures can be seen together in Plate 23 which shows a contact preparation of achiral and chiral smectic C phases. The racemic mixture of the two optically active isomers of 4-(2'-methylbutyl)phenyl 4'-n-octyloxy-biphenyl-4-carboxylate (top of plate) has been allowed to form a contact with (+)-4-(2'-methylbutyl)phenyl 4'-n-octyloxybiphenyl-4-carboxylate (lower part of plate).

(2) The petal texture

This texture is observed only when the helix of the chiral C phase is of a pitch such that iridescent light in the visible region of the spectrum is reflected. The texture is therefore a pseudo-homeotropic texture that selectively reflects coloured light. The texture is usually obtained by cooling a cholesteric phase (and sometimes a smectic A phase) and subjecting the preparation to mechanical stress, when the chiral C phase forms. The texture changes colour with temperature (Gray and McDonnell, 1976b); at lower temperatures the texture appears blue and at higher temperatures it appears red. Plate 24 shows the 'petal' texture of the chiral C phase of (+)-4-n-hexyloxyphenyl 4'-(4"-methylhexyl)-biphenyl-4-carboxylate.

(3) The focal-conic fan texture

The fan texture of the phase is usually obtained either by cooling a cholesteric or a smectic A phase. The fan backs are broken in a similar way to those of the achiral C phase, but the dark shaded areas are now coloured, thus giving the fan a shimmering appearance. The colour changes as the sample is rotated, in a similar way to that in which the dark areas would change for an achiral smectic C phase. Plate 69 of Sequence 2 shows the focal-conic fan texture of the chiral smectic C phase of (+)-4-(2'-methylbutyl)phenyl 4'-n-octyloxybiphenyl-4-carboxylate.

(4) The lined texture

This is a very special texture of the chiral C phase, and occurs when the layer planes dip at an angle to the surface of the supports. This produces a lined region similar to the Grandjean terraces observed in wedge-shaped samples of cholesterics. The lines are equally spaced and the width between them is directly related to the pitch length of the helix of the phase. Plate 25 shows the lined texture of the chiral smectic C phase of (+)-terephthalylidene-bis-4-(4'-methylhexyloxy)aniline.

Identification and classification of the smectic C phase

Microscopic textures

The following practical points may usefully be summarized.
(a) The ordinary smectic C phase usually exhibits two natural (spontaneous) textures, of which the focal-conic texture can also occur paramorphotically. The phase is characterized by being tilted, and therefore exhibits a biaxial interference figure. Of the two textures, the focal-conic fan texture is broken and sanded, thus giving it a rather ill-defined appearance. The *schlieren* texture is similar to that of the nematic phase but does not show the same degree of Brownian motion, and it does not flash when subjected to mechanical stress. The *schlieren* texture shows only singularities with four associated brushes, unlike the nematic phase which can also exhibit centres with two brushes. These textures characterize the C phase.
(b) The naturally formed focal-conic fan texture is not as lined or broken as that of the fan texture obtained paramorphotically. Therefore, any other tilted phases formed on cooling such a C phase will also exhibit fan textures which are not as broken as would be expected.
(c) The birefringence colours of the *schlieren* texture change with temperature as the tilt angle changes. Usually, the first-formed texture is a pale yellow-grey colour which turns yellow-brown to rust-blue on

cooling. However, for other more highly structured tilted phases, e.g. smectics G and H, the change in tilt angle is not as great; therefore less marked changes in colour with temperature are observed.

Miscibility studies

(a) The C phase has been shown to be very dependent on molecular structure, and in a similar way, the C phase of mixtures formed in miscibility studies are very dependent on the *actual* structures or on associations (e.g. by hydrogen bonding) that may occur between the components. Thus, mixed dimers can arise from $R'O.C_6H_4.CO.OH$ and $R''O.C_6H_4.CO.OH$. Therefore, it is important to select standard smectic C materials for miscibility studies very carefully because the transition temperatures marking the upper limit of the C phase often fall away with increasing percentage of either constituent of the binary mixture, thus forming a minimum in the transition line. If the minimum formed by the C phase is below the thermal stability region of an orthogonal phase, such as a B phase, there will be an injection of orthogonal properties in the mid-percentage region of the diagram of state. This quite often happens and interrupts the continuum of smectic C properties across the diagram. Thus, it is important to pre-select a standard material that has a similar molecular structure to that of the test material.

(b) Standard materials that exhibit the smectic C phase and are useful in miscibility studies can be listed as follows:

(i) $C_8H_{17}O-\phenyl-CO.OH$

4-n-Octyloxybenzoic acid

$I \rightarrow N \rightarrow S_C$

(N to S_C material).

(ii) $C_4H_9-\phenyl-N=CH-\phenyl-CH=N-\phenyl-C_4H_9$

Terephthalylidene-bis-4-n-butylaniline (TBBA)

$I \rightarrow N \rightarrow S_A \rightarrow S_C \rightarrow S_G \rightarrow S_H$

(high-temperature S_C phase with other tilted phases formed on cooling).

(iii) $C_8H_{17}O-\bigcirc-\bigcirc-CO.O-\bigcirc-OC_6H_{13}$

4-n-Hexyloxyphenyl 4'-n-octyloxybiphenyl-4-carboxylate

I→N→S_A→S_C→S_B

(tilted C phase in a predominantly orthogonal sequence).

(iv) $C_{10}H_{21}O-\bigcirc-\bigcirc-CO.OC_6H_{13}$

4-n-Hexyl 4'-n-decyloxybiphenyl-4-carboxylate

I→S_A→S_C

(similar A to C transition).

(v) $C_{10}H_{21}O-\bigcirc-CO.O-\bigcirc-OC_7H_{15}$

4-n-Heptyloxyphenyl 4-n-decyloxybenzoate

I→S_A→S_C

(simple A to C transition; a simpler compound to prepare than (iv)).

X-ray diffraction pattern

The X-ray diffraction pattern of a powder sample of a smectic C phase shows a sharp inner and a diffuse outer ring, because of the liquid-like ordering of the molecules within the layers, and is similar to that of the smectic A phase. Because of the tilt angle, layer spacings are smaller than the molecular length. The tilt angle can be determined either by measuring the angle between the most intense regions of the inner ring and the outer ring (for oriented samples), or by measuring the lamellar spacing directly from the diffraction pattern and then calculating the tilt angle, using the known length of the molecule in its most extended conformation.

Differential scanning calorimetry

Both DTA and DSC show that the nematic or S_A to S_C transition is extremely weak and is second-order in nature. The enthalpies associated with these transitions are often very small and the transitions may go undetected even with the most sophisticated instruments. The approximate enthalpy values obtained for these types of transition are of the order 0.1 kcal mol^{-1} or less, i.e. <1 kJ mol^{-1}.

References

Bennett, G.M., and Jones, B. (1939). *J. Chem. Soc.* 420.
Cabib, D., and Benguigui, L. (1977). *J. Phys. (Paris)* **38,** 419.
Clark, N.A., and Lagerwall, S.T. (1981). Proceedings of the Eighth International Liquid Crystal Conference, Kyoto, 30th June–4th July, 1980. Paper number I.29.
De Jeu, W.H. (1977). *J. Phys. (Paris)* **38,** 1265.
Dianoux, A.J., Heidemann, A., Volino, F., and Hervet, H. (1976). *Mol. Phys.* **35,** 1521.
Doucet, J., Levelut, A.-M., and Lambert, M. (1973). *Mol. Cryst. Liq. Cryst.* **24,** 317.
Goodby, J.W., Gray, G.W., and McDonnell, D.G. (1977). *Mol. Cryst. Liq. Cryst. Lett.* **34,** 183.
Goodby, J.W., and Gray, G.W. (1978). *Mol. Cryst. Liq. Cryst.* **48,** 127.
Gray, G.W., and Goodby, J.W. (1976a). *Mol. Cryst. Liq. Cryst.* **37,** 157.
Gray, G.W., and McDonnell, D.G. (1976b). *Mol. Cryst. Liq. Cryst.* **37,** 189.
Keller, P., Jugé, S., Liebert, L., and Strzelecki, L. (1976). *C. R. hebd. Séanc. Acad. Sci. (Paris), Ser. C,* **282,** 639.
Leadbetter, A.J. (1979). 'Structural studies of nematic, smectic A and smectic C phases.' In G.R. Luckhurst and G.W. Gray (eds.), *The Molecular Physics of Liquid Crystals,* Academic Press, London and New York, pp. 285–316.
Luckhurst, G.R., and Timimi, B.A. (1979). *Phys. Lett.* **75A,** 91.
McMillan, W.L. (1973). *Phys. Rev. A* **8,** 1921.
Meyer, R.B., Liebert, L., Strzelecki, L., and Keller, P. (1975). *J. Phys. (Paris) Lett.* **36,** L-69.
Moncton, D.E., Pindak, R., and Goodby, J.W. (1980). Proceedings of the Meeting of the American Physical Society, New York.
Nehring, J., and Saupe, A. (1972). *J. Chem. Soc., Faraday Trans. II* **68,** 1.
Priest, R.G. (1975). *J. Phys. (Paris)* **36,** 437.
Priest, R.G. (1976). *J. Chem. Phys.* **65,** 408.
Sackmann, H., and Demus, D. (1966). *Mol. Cryst. Liq. Cryst.* **2,** 81.
Saupe, A. (1973). *Mol. Cryst. Liq. Cryst.* **21,** 211.
Van der Meer, B.W., and Vertogen, G. (1979). *J. Phys. (Paris)* **40,** 222.
Wulf, A. (1975). *Phys. Rev. A* **11,** 365.

4 The smectic D phase

Introduction

Although not then recognized as such, the smectic D phase was first observed by Gray and co-workers (1957) amongst the various smectic polymorphic modifications exhibited by the longer chain members of the homologous series of 4'-n-alkoxy-3'-nitrobiphenyl-4-carboxylic acids, i.e.

$$C_nH_{2n+1}O-\underset{NO_2}{\underset{|}{C_6H_3}}-C_6H_4-CO.OH$$

where $n=16$ and 18.

The homologues with $n=16$ and 18 were first reported to give three distinct polymorphic smectic phases. At that particular period, this phenomenon was in itself interesting, and the behaviour of these acids was a topic of discussion between Gray and Sackmann in 1965.

Subsequently, these acids were investigated by Sackmann and his research group at Halle (Demus and co-workers, 1968; Diele and co-workers, 1972a,b). The following sequences of phase transition were reported:

$n=16$:

$$\text{Crystal} \xleftrightarrow{126.8°} S_C \xleftrightarrow{171.0°} S_D \xleftrightarrow{197.2°} S_A \xleftrightarrow{201.9°} \text{isotropic}$$

$n=18$:

$$\text{Crystal} \xleftrightarrow{124.6°} S_C \xleftrightarrow{158.9°} S_D \xleftrightarrow{195.0°} \text{isotropic}$$

In the hexadecyloxy compound, the S_D phase therefore exists intermediate on the temperature scale between the lamellar S_A and S_C phases (which have been discussed in previous chapters). In the case of the octadecyloxy compound, a S_C phase again exists below the S_D phase, but no S_A phase now occurs above it. The S_D phase is transformed directly into the isotropic liquid at 195°. In both cases, the S_D phase appeared to exist over considerable temperature ranges (26.2° when $n=16$, and 36.1° when

$n=18$), and the above sequences of phase transition seemed to be reversible on cooling. However, more will be said about the latter point when supercooling effects involving the S_A to S_D and isotropic to S_D transitions are discussed.

The S_C and S_A phases of the two acids under consideration appear to behave quite normally. The S_C phase exhibits either a *schlieren* texture or a sanded texture, but the S_A phase has a marked tendency to be homeotropic. Only quite rarely are transient focal-conic groups observed, usually at temperatures near to the S_A–I transition.

The unique feature of the S_D phase is explained in the following way. When the S_C phase of either acid is under observation between crossed polarizers, the birefringent *schlieren* or sanded texture is readily visible. However, at the S_C–S_D transition, the observer becomes aware of the development of completely black areas (with well-defined shapes) in the S_C texture. These regions develop and grow until the entire field of view is black. When the optically extinct S_D phase is submitted to pressure or stress by touching the cover-slip over the preparation, *no* flashing or birefringence effects are observed as in the case of optically extinct, homeotropic films of nematic or S_A phases. Indeed, from this point of view, the S_D phase could be mistaken for an isotropic liquid, except that it is highly viscous, and obviously is a considerably structured phase. The S_D phase was originally thought to be homeotropic (pseudo-isotropic) and to be too viscous to show any shear birefringence. Later however, it became clear that the S_D phase is in fact optically *isotropic,* not homeotropic; investigations by Pelzl (1969) and Pelzl and Sackmann (1971) showed quite definitely that the phase is not in any way anisotropic.

Structure of the smectic D phase

Returning to the way in which the optically isotropic areas of the S_D phase develop in the S_C phase, two points of importance must be made. Firstly, the black, isotropic areas of the S_D phase have quite distinctive shapes—rectangles, squares, rhombs, and hexagons—and secondly, these growing areas eventually coalesce to give a uniformly, optically isotropic field of view, without any grain boundaries.

These observations would indicate that the S_D phase has a cubic lattice, and that the geometrically regular shapes of the growth areas of S_D in the S_C phase correspond to sections of the cubic S_D lattice along particular lattice planes. The absence of grain boundaries is consistent with the entire S_D film corresponding to a large 'single crystal' domain.

The idea of a cubic model for the S_D phase was first developed in Sackmann's research group at Halle, and despite the difficulties in dealing with a fairly high-temperature phase exhibited by acids which are not free from decomposition at the elevated temperatures involved, some interesting X-ray studies of the S_D phase were carried out (Diele and co-workers,

1972b). The X-ray diffraction patterns of the S_C and S_A phases were quite normal—an inner ring (sharp) and an outer ring (diffuse). The S_D phase gave a very weak and blurred outer ring at 4.5 Å (Bragg angle $\approx 10°$), but the inner ring had changed and gave clear evidence for spot-like interferences (Bragg angle $\approx 1°$). In the case of suitable samples of the hexadecyloxy compound, the spot-like interferences formed a well-defined hexagonal pattern, although in other cases, the number and the distribution of the spots were different.

These authors concluded that a hexagonal fibre-structure was not reasonable. The existence of the hexagonal arrangement of spots and the diffuse outer ring (characteristic of disordered alkyl chains) suggested to them that the structure must be well ordered in parts, but have a liquid-like distribution in other parts.

Any form of lamellar structure was eliminated by the optical properties, and a body-centred cubic model was therefore proposed in which the aromatic parts of the molecules constitute spherical micelles which form the ordered cubic lattice, and the long C_{16} or C_{18} carbon chains have an irregular, liquid-like distribution, giving rise to the diffuse scattering at the larger Bragg angles. With certain assumptions made (indexing of the first reflection as 110), it was proposed that the lattice parameter a of the cubic cell was approximately 61 Å. The authors did not suggest how this lattice spacing was related to the molecular structure of the 4'-n-alkoxy-3'-nitro-biphenyl-4-carboxylic acids, and said that more detailed investigations would be necessary to confirm their speculations about the structure of the S_D phase.

Since this work there have been developments which are discussed below, but the structure of the S_D phase still remains something of an enigma; the existence of a cubic, *possibly* micellar phase intermediate on the temperature scale in the case of the hexadecyloxy compound between a lamellar S_C phase and a lamellar S_A phase is quite fascinating. If micellar in nature, this would require that the lamellar S_C phase breaks up and forms micelles, probably approximately spherical in shape and containing many molecules, and that these micelles then form the cubic lattice. At a higher temperature, the micelles must either break down and form the disordered isotropic liquid as for $n=18$, or in the case of $n=16$, the breakdown must result in the formation of a further lamellar phase (S_A). This interesting possibility has been discussed by Coates and Gray (1976) and by Gray and Winsor (1974a) who point out that analogies for the existence of cubic, isotropic phases intermediate between two other phases do occur in the field of lyotropic liquid crystals. In these cases the cubic, isotropic phases are amphiphilic micellar phases. These authors also point out a further relationship of the S_D phase—namely that to the optically isotropic plastic crystal phases (Gray and Winsor, 1974b) given by non-amphiphilic materials like camphor and adamantane.

Of the developments since the publications stemming from the Halle group, one of considerable interest concerns further X-ray studies by Tardieu and Billard (1976) of the S_D phases of the 4'-alkoxy-3'-nitrobiphenyl-4-carboxylic acids ($n=16$ and 18). This work further strengthens the analogy between the S_D phase and certain cubic, isotropic phases occurring in lyotropic systems, but in this case the latter phases are not micellar in type. Using a rotating sample—a technique which Tardieu had earlier used with great success in her studies with Luzzati of certain lipid systems—striking resemblances between the X-ray results for the cubic, non-amphiphilic S_D phase and the cubic, amphiphilic phases of certain lipid systems were found. Tardieu and Billard therefore suggested that the S_D phase may have a jointed-rod structure made up of short cylinders, not a micellar structure. To explain the optical isotropy of the S_D phase, interpenetrating or interwoven, three-dimensional jointed-rod structures with overall cubic symmetry would be required as indeed proposed earlier by Luzzati and Spegt (1966) and Tardieu and Luzzati (1970) for certain optically isotropic lipid phases which possess cubic symmetry (Fig. 4.1). For further information on such jointed-rod models in lyotropic systems, the reader is referred to Fontell (1974). Tardieu and Billard also re-indexed the reflections so that the first reflection became 211, giving a lattice parameter $a=102$ Å.

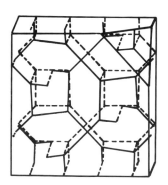 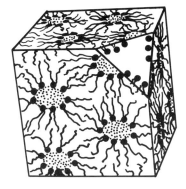

Figure 4.1 The jointed rod structure (with cubic symmetry) proposed by Luzzati and Spegt (1966) and Tardieu and Luzzati (1970) for certain optically isotropic lipid phases. Redrawn from Tardieu and Billard, *J.Phys. (Paris)* (1976).

One of the main factors inhibiting progress towards a clearer understanding of the cubic lattice of the S_D phase is unquestionably the limited number of compounds which exhibit this phase. As noted above, the C_{16} and C_{18} homologues of the 4'-alkoxy-3'-nitrobiphenyl-4-carboxylic acids were for many years the only two materials to form this phase. These materials are far from ideal for physical study because the S_D phase occurs at quite elevated temperatures ($>159°$) at which decomposition of the acids does occur. Maintenance of the purity of the materials in physical studies involving prolonged time intervals for the collection of the data is therefore a matter of concern.

Moreover, the nitro-acids synthesised according to the original procedure of Gray and co-workers (1957) are made by the nitration of the 4'-n-alkoxybiphenyl-4-carboxylic acids which are prepared from the 4-acetyl-4'-n-alkoxybiphenyls by hypobromite oxidation. In quite recent studies which have not been published, Gray and Mosley have shown that some ring-bromination can occur during the oxidation of the ketone, and that the resulting 4'-n-alkoxy-3'-bromobiphenyl-4-carboxylic acid can be carried through into the end product—the 4'-n-alkoxy-3'-nitrobiphenyl-4-carboxylic acid. Mass spectrometric analysis has revealed the presence of up to 5% of bromo-compound in some samples of the nitro-acid. Furthermore, this impurity is not easily removed by normal purification procedures. This does not necessarily mean that earlier work on the nitro-acids and their S_D phases has involved impure materials, but this possibility does exist.

As a development of this work, and as part of the quest for new S_D materials, Goodby and Gray (1980) have prepared the analogous cyano-compounds, i.e. the 4'-n-alkoxy-3'-cyanobiphenyl-4-carboxylic acids of structure

$$C_nH_{2n+1}O-\underset{\underset{CN}{|}}{\bigcirc}-\bigcirc-CO.OH$$

where $n=16$ and 18.

These new materials again exhibit S_D phases, and so they represent only the third and fourth examples of S_D systems.

It is worth noting that these cyano acids (and later the nitro acids) were prepared by a different route as shown schematically (Goodby and Gray, 1980) in Fig. 4.2. The unsubstituted acids ($R=C_{16}H_{33}$, $C_{18}H_{37}$) were brominated in the 3'-position, prior to replacement of the 3'-bromo-substituent by the 3'-cyano-substituent. Mass spectrometry confirmed that the new S_D materials were free from any bromo-compound.

This procedure now provides a better route to the 3'-nitro acids which can be prepared by direct nitration of a 4'-alkoxybiphenyl-4-carboxylic acid which cannot, because of its method of preparation, be contaminated by any bromo-isomer. This method of synthesis is recommended for obtaining materials for future physical studies of the S_D phases of these nitro-acids.

The S_D character of the cyano-acids seems to emphasize that the S_D phase requires a molecular structure involving a long terminal alkoxy chain and very strong lateral molecular interactions involving the dipoles

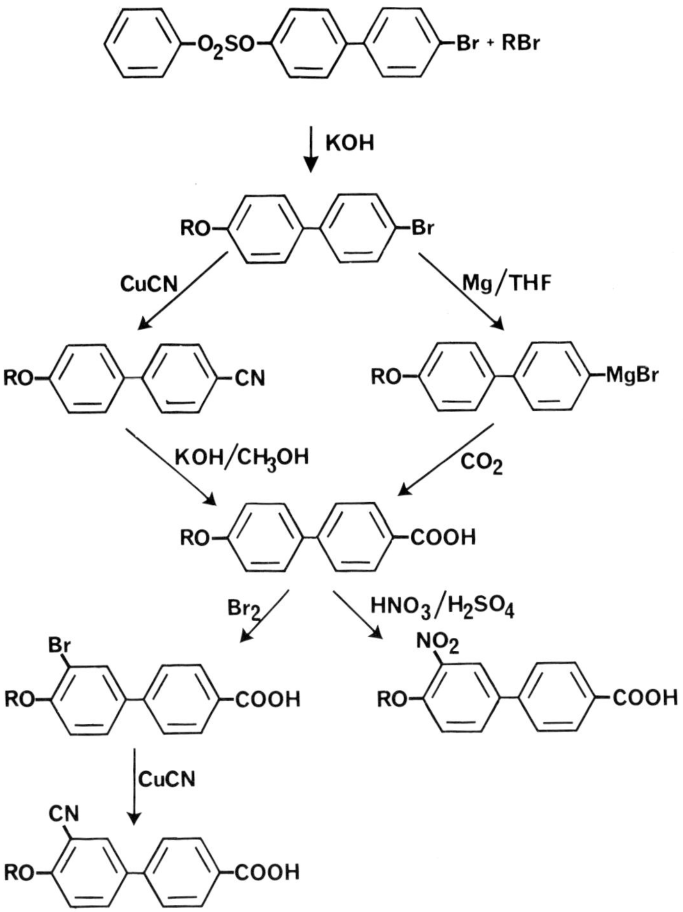

Figure 4.2 Synthetic route recommended for the preparation of S_D materials.

associated with either a lateral nitro- or a cyano-substituent.* The failure, until recently, to find the S_D phase in materials other than *acids* also seemed to be significant and to imply that a dimeric structure or some other intermolecular association involving hydrogen bonding of the carboxyl groups was also a prerequisite for formation of the phase.

Recently however, Demus (1981) has reported that cubic mesophases closely similar to those of the nitro- and cyano-acids occur in the 1,2-bis-(4-n-alkoxybenzoyl)-hydrazines with alkyl groups containing 8, 9, and 10 carbon atoms. These compounds were first reported by Schubert and co-workers (1978) as giving optically isotropic phases, possibly of the plastic crystal kind.

---

*Note that the analogous 4'-n-alkoxy-3'-halogenobiphenyl-4-carboxylic acids (halogen=F, Cl, Br or I) do not exhibit S_D phases (only S_C phases).

$C_nH_{2n+1}O-\bigcirc-CO.NH.NH.CO-\bigcirc-OC_nH_{2n+1}$

The phase sequence for these systems is

Crystal⟷cubic mesophase⟷S_C⟷isotropic

More will be said about these compounds later, but in the present context, it should be noted that although they are structurally different from the nitro- and cyano-acids, the molecules still retain (a) terminal alkoxy groups with fairly long chains and (b) a 'double' structure (arising from the hydrazine linkage) which is similar to that which exists in the dimers of the acids. Also, Demus points out that although the lateral, dipolar nitro- or cyano-groups are missing, strong lateral intermolecular interactions may again arise from hydrogen bonding interactions between adjacent di-hydrazine linkages.

None of the proposed cubic models takes these particular features of the molecular structure into account. Of course, the statistical situation is very weak with so few known S_D materials, and only time will reveal the importance, if any, of the long terminal alkoxy chains, the 'double' structure, the strong lateral interactions, and the aromatic core structure. Needless to say, attempts still go on to make structurally modified S_D systems, but the evidence is often of a negative nature, e.g. esterification of the carboxyl group of the nitro-acids (giving simple n-alkyl esters) produces simple S_A materials.

A further development of some interest concerns observations made by Demus and co-workers (1980) on the reversibility of the transitions given by the two nitro-acids, and the effect of cooling rate on the reversal. Their results have shown that the nucleation of the S_D phase in either the homeotropic texture of the S_A phase of the acid with $n=16$ or the isotropic liquid of the acid with $n=18$ may be markedly inhibited, whereon a metastable phase called S_4 may form; this was previously reported as S_D by Demus and Richter (1978). On cooling, this phase may then change into the isotropic S_D phase or, dependent on the degree of supercooling, remain until a sufficiently low temperature is reached for the formation of the S_C phase.

The situation they describe (showing the limits of supercooling and superheating) is summarized below (Demus, 1981).*

$n=16$:

$$\text{Crystal} \xrightarrow{126.8°} S_C \xrightarrow[-10°]{171° \quad +3°} S_D \xrightarrow[-30°]{198° \quad +0.4°} S_A \xrightarrow{199.8°} \text{isotropic}$$

with $170° \pm 2°$ and $191° \pm 2°$ and $193.3°$ going to S_4.

---

*Note that some minor differences in transition temperatures occur relative to the values quoted by Diele and co-workers (1972b).

$n = 18$:

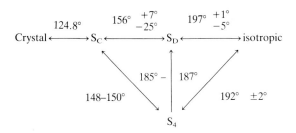

In the first edition of the book *Textures of Liquid Crystals* by Demus and Richter (1978), Plates 145 and 146 illustrate the texture of what they later refer to as the metastable S_4 phase. For the acid with $n=16$, Plate 145 shows what the authors describe as the commencement of the growth of birefringent mosaic grains in the homeotropic S_A phase. The temperature is 195° and Plate 146 shows the completion of the transition at 193°, when the whole field of view is birefringent. The mosaic grains are also described in the text as bâtonnets (although they are not typical of these) which ultimately form what is described as a mosaic texture. At the time the book was written, prior to the article by Demus and co-workers (1980), the authors say the following:

> This mosaic texture is existent for a short time only and changes spontaneously into the isotropic texture at lower temperatures. The mosaic texture is not compatible with the cubic structure of smectic D. It has to be assumed that, in a first step, the planar layer structure [of S_A] alters to a structure with layers or bands, whereas in a second step, it transforms into the cubic structure with micellar units.

As a result of their later work, it was concluded by Demus and co-workers (1980) that the birefringent texture is not S_D, but a texture of a new metastable S_4 phase which does not have a cubic structure and can transform under suitable conditions of temperature into S_C or S_D or S_A. That is, both supercooling and superheating effects arise and lead to the new metastable S_4 phase.

More recently, Dr J.E. Lydon (1981) has drawn together the factual information on the S_D phase of the nitro- and cyano-acids and produced a self-consistent picture of its structure, of which he has kindly permitted the authors to give the following summary. Lydon sees a very close similarity between the texture reproduced by Demus and Richter (1978) for the S_4 phase and that observed for the discotic phase of di-isobutylsilanediol (Bunning and co-workers, 1980, 1981) and by many other workers for the discotic phases of other materials whose molecules have a disc-like shape. Indeed, at the S_A–S_4 transition, the textural similarity is remarkably striking.

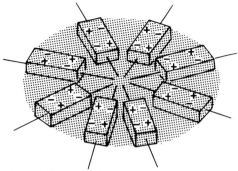

Figure 4.3 Aggregation of dimeric molecules into discs which it is proposed form the columns of which the hexagonal arrangement gives the S_4 phase. Figures 4.3, 4.4 and 4.5 reproduced from Lydon, *Mol. Cryst. Liq. Cryst.* (1981) by permission of Gordon and Breach Science Publishers.

It is proposed that the principal structural requirement for formation of the S_D and the S_4 phase is the presence of the strong lateral dipoles. The carboxyl groups are considered to form the hydrogen-bonded dimer, and a plausible arrangement of the molecules is that shown in Fig. 4.3. That is the molecules may aggregate to form discs. As in discotic phases, there will be a tendency for these discs to stack up and form columns or cylinders which then arrange themselves in a hexagonal array. Consequently, the S_4 phase is considered to be a discotic phase.

Lydon however introduces a further argument. In the cylinders, it will be likely that the terminal alkyl chains *in the centre* do not have as much room as they need, and they will tend to protrude, giving the cylinders pointed ends (Fig. 4.4). He proposes that as the cylinder grows in length, the vertex angle increases until it becomes 120°. At this stage, and only at this stage, the units can pack together and form an infinite lattice of the type suggested by Tardieu and Billard (1976)—see Fig. 4.5. For this lattice to form, all the cylinders need to be the same length, and this situation would be assured by the relationship between the length of the cylinder and the vertex angle.

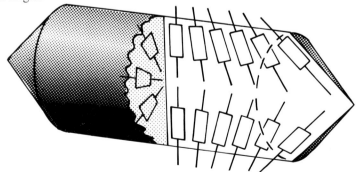

Figure 4.4 Molecular aggregation in the S_D phase. Cylinders with pointed ends, which form the jointed rod structure of Tardieu and Billard—see Fig. 4.5.

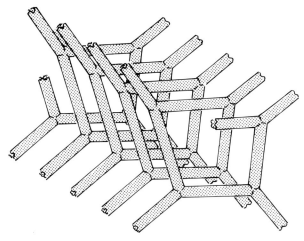

Figure 4.5 Drawing of part of the structure of the jointed rod model of Tardieu and Billard for the S_D phase. Units of this type are linked together to form two interwoven networks with an overall cubic symmetry.

Therefore, as the temperature falls from either the isotropic liquid or the S_A phase, the disc-like units could form either the discotic phase or the S_D phase, dependent upon the extent of supercooling. The discotic phase could also rearrange either spontaneously or under stress (when the cylinders become large enough) to give the cubic D phase, or if sufficient supercooling occurs, the discotic phase could transform directly to give the S_C phase.

The original Tardieu and Billard model did not incorporate dimerization of the molecules, and these authors suggested that their proposed structure was compatible with either the alkyl chains or the carboxyl groups pointing towards the cylinder centres. As Lydon remarks, this leads to awkward volumes to be filled by non-flexible aromatic parts of the molecules. Adopting the dimer avoids these problems but gives a lattice of twice the size of that of Tardieu and Billard. Quoting from Lydon, '. . . it might be thought that the X-ray diffraction data effectively rule out this possibility. However, it is to be expected that a structure of the type proposed will give a systematic pattern of strong and weak reflections and it is feasible that this causes an apparent halving of the lattice dimensions.'

Finally, we must return to the recent article by Demus and co-workers (1981) in which the similarities between the cubic isotropic mesophases of the nitro-acids and the hydrazine derivatives are emphasized. Not only are the X-ray diffraction patterns similar, but also the phase transition from S_C to the cubic mesophase supercools strongly. However, it is stated that the cubic mesophases of the nitro-acids and those of the hydrazine derivatives are *not* continuously miscible with one another. This suggests a polymorphism of this type of phase. According to the miscibility rules of Sackmann and Demus, the code letter S_D could not therefore be applied to both phases. This problem is however avoided by Demus. Since he considers that these cubic phases are micellar, and not lamellar in nature, he notes that it is not really appropriate to classify them as of the smectic type. Indeed, in his article he no longer uses the term S_D, and refers to the phases simply as cubic mesophases. He also drops the term smectic from his description of the phase he originally described as S_4, and now calls this phase (considered by Lydon to be discotic) mesophase 4.

It is of interest to note that Demus (1981) concludes his paper (written it should be emphasized with no knowledge of Lydon's ideas about discotic/S_D inter-relationships) by noting that these cubic phases formerly designated S_D appear to constitute a third system* of thermotropic mesophases, the other two being the systems embracing (a) nematic and smectic mesophases, and (b) discotic mesophases.

In the remainder of this chapter, we will however continue to use the terms S_D and S_4 to describe these phases—where S_D is the cubic isotropic phase with either a micellar or a rod structure, and S_4 is the phase which is most probably discotic.

Textures of the smectic D phase

Since the S_D phase is optically isotropic, it has no texture in the real sense. A uniform film of S_D presents an optically extinct (black) field of view. The phase is distinguished from the isotropic liquid or the homeotropic phases by the fact that the black, isotropic S_D phase nucleates in the birefringent S_C phase (occurring at lower temperatures) in straight-edged squares, rhombs, hexagons, and rectangles (Fig. 4.6(a) and Plate 26). The very well developed areas in Plates 147 and 148 of Demus and Richter (1978) are somewhat exceptional. Sometimes the D phase is nucleated in fern-type patterns which probably consist of a number of cubic domains (probably octahedral) growing together (without detectable grain boundaries) as shown in Plates 27 and 28 and Fig. 4.6(b). Although the black fern texture growing in the *schlieren* S_C is optically isotropic, it bears some resemblance to the fern patterns observed for crystal B phases nucleating from the nematic phase (compare with Plate 16).

When the S_D phase transforms to or nucleates from either the isotropic liquid or the S_A phase in its homeotropic texture, the transitions are hard to observe—uniform black to uniform black. However, detection can be facilitated by (i) mechanical displacement of the cover-slip, (ii) observation of the effects of stress on any air bubbles in the thin film and the

---

*In which two immiscible polymorphic forms already appear to exist.

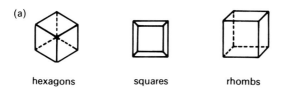

(a) hexagons squares rhombs

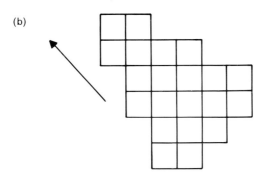

Figure 4.6 (a) Textural patterns for the S_D phase.
(b) Dendritic growth pattern of the S_D phase formed from the S_C phase on heating—compare with Plate 27.

accompanying motion of any dust particles in the melt. These observations can be aided by the use of a quarter-wave plate.

With very slow cooling rates (0.2°/min), the isotropic or S_A to S_D transition often supercools with formation of the S_4 phase. In some cases the S_4 phase develops as quite strongly birefringent rod structures which Demus and Richter (1978) regard as nucleation in the form of bâtonnets; this results (they say) in a mosaic texture. The S_4 phase was also found to form the isotropic D texture spontaneously.

Alternatively, the S_4 phase formed has rod-like structures set in a homeotropic texture, the phase being weakly birefringent (see Plate 29). This texture is similar to that exhibited by the discotic phase of di-isobutylsilanediol (compare with Plate 121). Perturbation of this texture spontaneously converts it to the isotropic texture of the D phase.

At the lower end of the temperature range of the S_4 phase, the transformation to the C phase is often, but not always, accompanied by a transition to the D phase first. Furthermore, as the S_4 or discotic texture can change spontaneously or with mechanical stress into the isotropic, cubic S_D texture, this latter texture must therefore be regarded as the natural texture of the D phase.

Identification and classification of the smectic D phase

Miscibility studies

(a) The miscibility situation is as follows. The S_D phases of all four S_D materials are continuously miscible over the entire composition range of all the possible binary mixtures that may be examined. Conversely, the S_D phases of these materials are totally immiscible with all other smectic phases so far assigned code letters. However, co-miscibility of the S_4 phase with a discotic phase of a true discotic material has not yet been investigated. Moreover, as already mentioned, the cubic phase of the hydrazine derivatives is not miscible with that of the nitro-acids.

(b) Standard materials that exhibit the original S_D phase are restricted to four, of which the most useful is probably 4'-n-hexadecyloxy-3'-nitro-biphenyl-4-carboxylic acid, since the S_D phase occurs in addition to a S_C and a S_A phase.

$$n\text{-}C_{16}H_{33}O-\underset{}{\underset{NO_2}{\bigcirc}}-\bigcirc-CO.OH$$

$$I \longleftrightarrow S_A \longleftrightarrow S_D \longleftrightarrow S_C$$
$$\uparrow \quad \uparrow \quad \uparrow$$
$$\rightarrow \quad S_4 \quad \leftarrow$$

This material also forms the S_4 discotic phase from the homeotropic S_A phase when supercooling of the S_A–S_D transition occurs. S_4 may either revert spontaneously to S_D, or if supercooling occurs, the S_C phase is produced at a temperature close to or just below the S_D–S_C transition.

X-ray diffraction pattern

The diffraction pattern consists of a diffuse outer ring at 4.5 Å characteristic of disordered alkyl chains. At small Bragg angles, interference spots occur which may, dependent on alignment, assume a hexagonal arrangement derived from the cubic lattice of the isotropic D phase.

Enthalpy data

Reliable enthalpy data on transitions involving the cubic (S_D) and discotic (S_4) phases do not appear to be available.

References

Bunning, J.D., Goodby, J.W., Gray, G.W., and Lydon, J.E. (1980). 'The classification of the mesophase of di-i-butylsilanediol.' In W. Helfrich and G. Heppke (eds.), *Liquid Crystals of One- and Two-Dimensional Order, Springer Series in Chemical Physics 11,* Springer-Verlag, Berlin, Heidelberg, and New York, pp. 397–402.

Bunning, J.D., Lydon, J.E., Eaborn, C., Jackson, P.M., Goodby, J.W., and Gray, G.W. (1981) *J. Chem. Soc., Faraday Trans. I* **78**, 713.

Coates, D., and Gray, G.W. (1976). *The Microscope* **24**, 117.

Demus, D., Kunicke, G., Neelson, J., and Sackmann, H. (1968). *Z. Naturforsch.* **23a**, 84.

Demus, D., and Richter, L. (1978). 'The textures of smectic D.' In *Textures of Liquid Crystals,* V.E.B. Deutscher Verlag für Grundstoffindustrie, Leipzig. Chap. 4.10, pp. 91–92.

Demus, D., Marzotko, D., Sharma, N.K., and Wiegeleben, A. (1980). *Krist. und Tech.* **15**, 331.

Demus, D. (1981). Thermotrope kubische Mesophasen. Presented at the Freiburger Arbeitstagung Flüssigkristalle, paper no. 1.

Demus, D., Gloza, A., Hartung, H., Hauser, A., Rapthel, I., and Wiegeleben, A. (1981). *Krist. und Techn.* **16**, 1445.

Diele, S., Brand, P., and Sackmann, H. (1972a). *Mol. Cryst. Liq. Cryst.* **17**, 84.

Diele, S., Brand, P., and Sackmann, H. (1972b). *Mol. Cryst. Liq. Cryst.* **17**, 163.

Fontell, K. (1974). 'X-ray diffraction by liquid crystals—amphiphilic systems.' In G.W. Gray and P.A. Winsor (eds.), *Liquid Crystals and Plastic Crystals,* Vol. 2, Ellis Horwood, Chichester, Chap. 4, pp. 80–109.

Goodby, J.W., and Gray, G.W. (1980). Unpublished results.

Gray, G.W., Jones, B., and Marson, F. (1957). *J. Chem. Soc.,* 393.

Gray, G.W., and Winsor, P.A. (1974a). *Mol. Cryst. Liq. Cryst.* **26**, 305.

Gray, G.W., and Winsor, P.A. (1974b). 'Classification and organization of mesomorphous phases formed by non-amphiphilic and amphiphilic compounds.' In G.W. Gray and P.A. Winsor (eds.), *Liquid Crystals and Plastic Crystals.* Vol 1, Ellis Horwood, Chichester, Chap. 2, pp. 18–63.

Luzzati, V., and Spegt, P.A. (1966). *Nature* **210**, 1351.

Lydon, J.E. (1981). *Mol. Cryst. Liq. Cryst. Lett.* **72**, 79.

Pelzl, G. (1969). Dissertation, University of Halle.

Pelzl, G., and Sackmann, H. (1971). *Symposium of the Chem. Soc., Faraday Division* **5**, 68.

Schubert, H., Hauschild, J., Demus, D., and Hoffmann, S. (1978). *Z. Chem.* **18**, 256.

Tardieu, A., and Luzzati, V. (1970). *Biochim. biophys. Acta* **219**, 11.

Tardieu, A., and Billard, J. (1976). *J. Phys. (Paris)* **37**, 79.

5 The smectic E phase

Introduction

This highly ordered, orthogonal smectic phase was initially discovered by Kölz (1966), and later reported on by Diele, Brand, and Sackmann (1972) and by Demus and Sackmann (1973) of the Halle group. The original reports concerning this new phase were made in relation to two compounds:

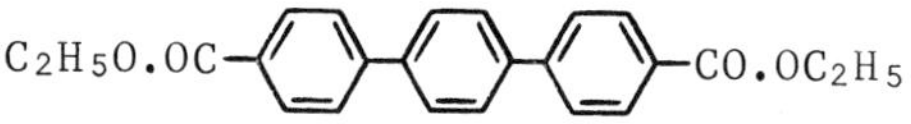

$C_2H_5O.OC$—⟨⟩—⟨⟩—⟨⟩—$CO.OC_2H_5$

diethyl *p*-terphenyl-4,4″-dicarboxylate

and

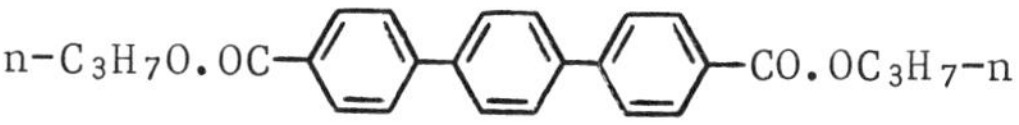

n-$C_3H_7O.OC$—⟨⟩—⟨⟩—⟨⟩—$CO.OC_3H_7$-n

di-n-propyl *p*-terphenyl-4,4″-dicarboxylate

These two compounds were found to exhibit a new smectic phase which was reported as being uniaxial. At a similar time to these reports, Gray and Harrison (1971a,b) and Coates, Gray, and Harrison (1973) made disclosures of another type of compound which exhibited a phase that had very similar microscopic textures to those of the phase reported by the Halle group. These compounds were of the cinnamate ester type shown below:

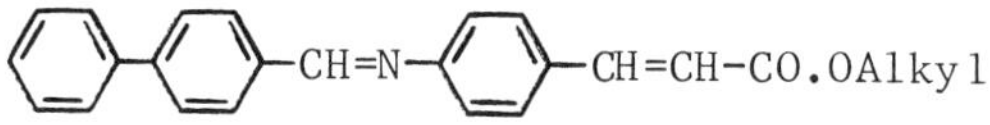

⟨⟩—⟨⟩—CH=N—⟨⟩—CH=CH-CO.OAlkyl

This new phase was however unlike the uniaxial phase in that it exhibited positive biaxiality. However, the two phases were shown to be co-miscible, and therefore, were of the same miscibility group. This miscibility group was shown to be novel and it was given the code letter E. The two variations of the phase then became known as the uniaxial smectic E phase and the biaxial smectic E phase.

All subsequent discoveries of new materials that exhibit this type of phase have been found to involve biaxial properties. Thus, the only examples of the uniaxial phase were the two terphenyl diesters mentioned above. However, considerably later, investigations of di-n-propyl p-terphenyl-4,4″-dicarboxylate by Goodby and Gray (1979) revealed that the S_E phase was also of the biaxial type. They also found that slight impurities were capable of producing textures for the E phase which appeared to be homeotropic and gave the impression that the phase was uniaxial. Therefore, it seemed likely that the S_E phase of the diethyl ester would also be of the biaxial type, although it was not included in their investigation, and the general conclusion that could be reached was that the S_E phase is biaxial, even though it is of the orthogonal type.

Structure of the smectic E phase

The smectic E phase has been the subject of numerous structural studies by X-ray diffraction techniques. The early studies by Diele, Brand, and Sackmann (1972) showed that the E modification (in comparison with the A and C phases, for example) has a higher order of the molecules within the layers. Furthermore, they concluded that in the E phase the indications were that the order went beyond that within single layers.

In a classical single-crystal X-ray diffraction study of the S_E phase, Levelut, Doucet, and Lambert (1974) were able to show that the molecules were in an orthorhombic array within the layers (Fig. 5.1).

In Fig. 5.1, the cross-sectional areas of the molecules at right angles to the molecular long axes are represented as circles. That is, at this point we are still assuming (but see later) that the molecules can rotate through 2π, if need be co-operatively, about their long axes. We can see therefore that the difference between a S_B and a S_E phase or the change that occurs when S_B changes to S_E (as is often the case on cooling) involves a contraction of the hexagonal net in one of three possible directions separated by angles of 60°. This contraction and the resulting orthorhombic arrangement account for the biaxial properties of the phase.

Further single-crystal X-ray diffraction studies by Doucet and co-workers (1975) confirmed that the biaxiality of the smectic E phase was not connected with a tilted arrangement of the molecules. The lamellar spacing was found to be close to that of the molecular length, and thus the molecules must be orthogonal to the layer planes (Fig. 5.2).

SMECTIC LIQUID CRYSTALS

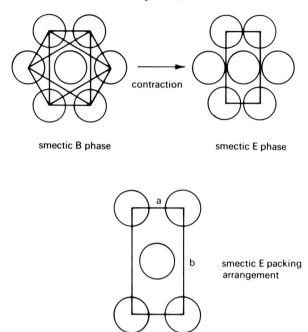

Figure 5.1 Distribution of the rod-like molecules within the layers of the smectic B and smectic E phases.

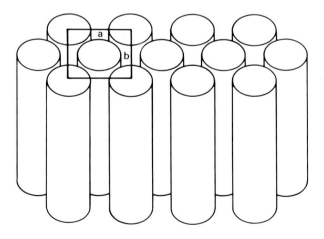

Figure 5.2 The packing arrangement of molecules in a layer in the smectic E phase ($b>a$)—see also Fig. 5.1.

Whereas for the smectic B phase, Levelut, Doucet, and Lambert (1974) were able to show that the unit cell dimensions were just large enough to permit rotation, albeit co-operative, of the molecules about their long axes, the dimensions for the unit cell of the smectic E phase (in relation to the dimensions of the molecule under investigation) indicated that free or even co-operative rotation was simply not possible. Thus, the problem of accommodating the molecules in an orthorhombic array, whilst still allowing them some rotational motion, and that of the biaxiality of the phase were explained in terms of a chevron-like arrangement of the cross-sectional areas of the molecules (Fig. 5.3). The elliptical shapes in Fig. 5.3 represent the cross-sections of the molecules looking directly down the long molecular axes.

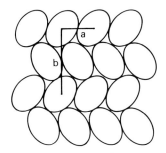

Figure 5.3 Chevron or herringbone packing of the molecules in the smectic E phase.

In this close-packed arrangement, the molecules are allowed to oscillate through angles of <180° (Fig. 5.4), and occasionally through 360° for high-energy molecules. In Fig. 5.4, we represent the cross-sections of the blade-like molecules simply as lines.

The dynamics of the molecular motion in the smectic E phase were the subject of exhaustive studies by Leadbetter, Richardson, and Carlile (1976) using X-ray diffraction and neutron scattering techniques—see also Leadbetter and co-workers (1979a, b). From their detailed experiments, they were able to show that the molecules were still moving about their long axes quite rapidly. However, X-ray data indicated that the packing arrangement of the molecules was too close for the rotation of the molecules about their long axes to be free. The experimental results were interpreted in terms of the oscillation of the molecules through a restricted angle of less than 180°. This gives the impression of a flapping motion of the molecules. This flapping reorientational motion is rapid ($\sim 10^{11}$ s^{-1}), but the molecules can still be regarded as spending the greater proportion of their time in the biaxial, two-fold disordered, chevron arrangement. However, as already noted, a molecule with sufficient energy can still rotate fully through 360°.

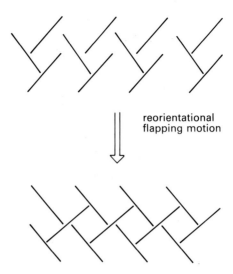

Figure 5.4 Oscillatory reorientational rotation of the molecules about their long axes in the smectic E phase.

The difference between this rotational situation in the S_E phase and that in the S_B phase, where rotation is still restricted, has been well summarized by Doucet (1979). 'The S_B phase is characterised by rotational disorder of the molecules around their long axes in such a way that on average, all the molecular positions are equivalent and that locally the lateral molecular arrangement is a herringbone packing with three orientations of the local domains.' He then goes on to say: 'In the S_E phase, the herringbone packing extends over long distances; the presence of two perpendicular glide mirrors implies a disorder involving a two-fold axis of reorientation through an angle π.' The emphasis therefore lies on local herringbone order in S_B and long-range herringbone order in S_E.

Later X-ray diffraction studies of the S_E phase have been carried out, particularly by Leadbetter and his colleagues (Richardson, Leadbetter, and Frost, 1978; Leadbetter, Frost, Gaughan, and Mazid, 1979a; Leadbetter, Frost, and Mazid, 1979b). From these, it has become apparent that the layers in the smectic E phase are correlated with each other. This gives rise to a three-dimensional crystal-like structure. The correlation length between layers is usually assumed to be infinite, but more realistically it is probably of the order of 300 to 1000 layers. This gives rise to a structure of the smectic E phase of the kind represented in Fig. 5.5, where the molecules are represented as elongated units with an elliptical cross-section.

THE SMECTIC E PHASE

Figure 5.5 The layer correlated smectic E phase.

As a part of their extensive X-ray studies of the S_E phase, Leadbetter and his colleagues made a particularly detailed study of isobutyl 4-(4′-phenylbenzylideneamino)cinnamate (IBPBAC)

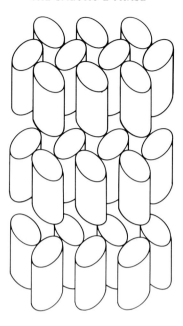

—not only of its smectic phases, but also of the structure of the solid crystal (Leadbetter, Mazid, and Malik, 1980).

The crystal structure (monoclinic) is a bilayer with the molecules tilted at ~40° to the layer normal and packed in each half-layer in a distorted hexagonal arrangement.

The relationship between this crystal structure and the structure of the S_E phase to which the solid gives rise on heating to 86° has been discussed. This work has revealed that subtleties of structure may occur in S_E phases, and this is not surprising when it is remembered that in the E phase we are dealing with a highly ordered crystal smectic phase. In the work of Leadbetter and his colleagues, the aim has been to carry out the X-ray studies on as well aligned samples of the S_E and S_B phases as possible. Thus, the subtle differences in structure that they detect would not have been discerned in earlier X-ray studies which employed powder samples or less well aligned mono-domains. Even so, they could not achieve *true* mono-domains and so the interpretation of their results is not unambiguous, and there is more than one possibility. Their arguments are detailed, but their broad conclusion is that for IBPBAC, although the packing of the molecules within the layers is confirmed as respectively distorted and true hexagonal in the S_E and S_B phases, the S_E structure is probably an orthorhombic bilayer, but it could be monoclinic with a very small tilt angle of ~5°. For the S_B phase, the structure is locally monoclinic with a 6° tilt angle and poor interlayer correlation, resulting in an overall uniaxial structure.

These results, whilst agreeing broadly with the main structural considerations given earlier in this chapter for the S_E phase, do show that detail of the generic structure of S_E phases is an open question. More work must still be done to find whether the features found for IBPBAC are in some way special or are common to all S_E compounds.

Textures of the smectic E phase

The smectic E phase has been obtained both on direct cooling of the isotropic liquid and on cooling smectic A or smectic B phases. Thus, the phase exhibits a natural microscopic texture and a number of paramorphotic textures inherited from precursor smectic phases.

The natural texture of the E phase

The number of compounds which exhibit direct smectic E to isotropic liquid transitions is very few. They are mainly limited to compounds of the type

$$C_nH_{2n+1}O-\!\!\left\langle\!\bigcirc\!\right\rangle\!\!-\!\!\left\langle\!\bigcirc\!\right\rangle\!-COC_mH_{2m+1}$$

where n and m are usually limited to values $n<12$ and $m<3$.

An example is 4-ethoxy-4'-acetylbiphenyl:

$$C_2H_5O-\!\!\left\langle\!\bigcirc\!\right\rangle\!\!-\!\!\left\langle\!\bigcirc\!\right\rangle\!-COCH_3$$

$$C \xrightarrow{96°} S_E \xrightarrow{156.2°} I$$

This small group of materials appears to represent all of the known mesogens giving S_E to isotropic liquid transitions, and hence the variety of natural textures observed for the S_E phase is limited. Basically, the natural texture of the smectic E phase separates from the isotropic liquid in the form of a droplet, as shown in Plate 30 for 4-ethoxy-4′-acetylbiphenyl. The droplets grow rapidly on cooling, forming an undulating, mosaic-type texture, as shown in Plate 31 for 4-ethoxy-4′-acetylbiphenyl. This texture formed by the E phase when it nucleates directly from the isotropic liquid has not as yet been fully investigated and its true nature remains uncertain at the present moment.

The paramorphotic textures of the smectic E phase

There are basically three types of texture exhibited by the smectic E phase that are derived from the textures of preceding smectic phases. These are the focal-conic fan texture, the platelet texture, and the mosaic texture. Other textures are however obtained by cooling a B phase that has formed via a I–S_{AB} transition.

The focal-conic fan texture is obtained on formation of the E phase from the focal-conic fan textures of either the smectic A or the smectic B phase. At the point of transition to the E phase, the focal-conic fans of the preceding phase become crossed with concentric lines or arcs running across the backs of the fans. The arcs run parallel to the curved parallel layers where these meet the surface of the preparation on which the microscope is focused—see section on the focal-conic fan structure of the smectic A phase. The arcs are not transitory in nature like the transition bars formed at a S_A to S_B transition, but are permanent and remain throughout the temperature range in which the phase persists. The paramorphotic arced focal-conic fan texture of the smectic E phase obtained on cooling the focal-conic textures of the S_A and S_B phases of methyl 4′-n-octyloxybiphenyl-4-carboxylate is shown in Plate 32. The arcs are caused by the different contraction directions of the layer ordering at the transition to the E phase (similarly to Fig. 5.6).

The platelet and mosaic textures are the other two paramorphotic textures exhibited by the E phase. The platelet texture is usually obtained when the smectic E phase is formed on cooling the homeotropic texture of a smectic B or a smectic A phase. At the transition from either of these phases to the E phase, the black, homeotropic area starts to show birefringent platelet areas. These platelets usually develop rapidly, and they often have a hexagonal or rhomb-like shape. The platelets grow and overlap one another as the phase transition becomes complete. Plate 33 shows the platelet texture of the smectic E phase formed on cooling the homeotropic texture of the smectic A phase of di-n-propyl p-terphenyl-4,4″-carboxylate.

The mosaic texture of the smectic E phase is usually obtained on cooling the mosaic texture of a B phase, which in turn has been obtained by forming the smectic B phase directly from the isotropic liquid. In this case,

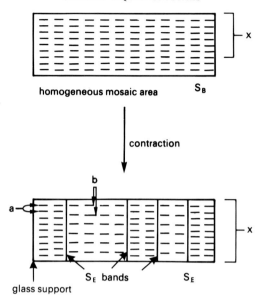

Figure 5.6 Origin of the black lines in the mosaic areas of S_E. The top part of this diagram represents a view from *above* (as seen under a microscope) of an idealized region of a *single* homogeneous S_B mosaic area, i.e. the layers pass downward into the preparation at 90° or some other angle to the plane of the page—see Fig. 2.12(*a*)(*b*) of Chapter 2. Contractions in the hexagonal net, e.g. as shown in the lower part of the diagram, could then occur in relation to stacks of correlated layers giving dislocations (black lines) in the texture.

the B phase produces its natural mosaic texture, and on cooling, as the E phase forms, the mosaic platelets become crossed with parallel lines. These lines are formed as the transition occurs and remain throughout the temperature range of the phase. These parallel lines probably have similar origins to those of the concentric arcs formed in the paramorphotic focal-conic fan texture. The lines may be rationalized in terms of contractions of the hexagonal S_B net in different directions in relation to stacks of correlated layers as shown in Fig. 5.6.

Finally, the E phase can show two further more specialized paramorphotic textures when it is obtained by cooling a smectic B phase that has formed from the isotropic liquid *via* an extremely short smectic A phase, i.e. I→S_{AB}→S_E. In this case the 'fans' formed in the B phase are usually aggregated in rows giving an elongated domain. On cooling to the E phase, this domain becomes crossed with a number of parallel lines. These lines usually run parallel to the longer edge of the domain, as shown in Plate 34 for the smectic E phase of 3-methylbenzyl 4-(4'-phenylbenzylideneamino)cinnamate.

When the B phase formed *via* a I–S$_{AB}$ transition exhibits a homeotropic texture, then on cooling to the E phase, the texture becomes birefringent and forms moss-like areas—see Plate 34, i.e. in this plate the mossy regions were homeotropic in the B phase. There is no reasonable explanation of the structure of this texture at the present moment.

Identification and classification of the smectic E phase

Microscopic textures

(a) The E phase is usually obtained by cooling either the A or the B phase. It is not normally observed in a phase sequence that contains ordered, tilted smectic phases. The phase can be obtained in some cases by direct nucleation from the isotropic liquid; however, this is usually a rare occurrence.

(b) The smectic E phase can be readily identified from its arced paramorphotic focal-conic fan texture. The lines of the arcs are usually very clear and the bands (areas between the lines) are unbroken. This texture is very characteristic of the E phase; no other phase shows this type of texture, except the smectic G phases formed on cooling smectic B phases. In this case, the arcs are seen, but they are dissimilar to those of the E phase because they are broken, i.e. not continuous throughout the preparation as they are in the E phase.

(c) The E phase exhibits a platelet texture which often appears grey-blue to yellowish in colour. The texture is very different to normal mosaic textures in that the transparent platelets overlap so that ghost-like images of platelets can be seen through platelets near to the surface. This type of texture is unique to the E phase and therefore makes it readily identifiable.

Miscibility studies

(a) Miscibility studies involving smectic E materials are often difficult to perform. The major problem lies with the binary mixtures containing approximately 40 to 60% of one component. Often the smectic E thermal stability decreases in this region, giving a depression in the transition temperature curve for the S_E to S_A or S_B transitions. If this depression falls below the line through points connecting the recrystallization temperatures of the various mixtures, then it is impossible to establish continuous miscibility of the E phase across the diagram of state. Thus, for the purpose of miscibility studies, it is advisable to select a standard material which has a similar molecular structure to that of the test compound, since the extent of the depression is then minimized. However, it should be noted that even using mixtures of members of the same homologous series, large depressions in the E phase thermal stability can occur.

(b) Standard materials that exhibit the smectic E phase and are useful in miscibility studies:

(i) $C_2H_5O-\langle\rangle-\langle\rangle-COCH_3$

4-Ethoxy-4'-acetylbiphenyl (S_E to isotropic liquid transition).
$$I \rightarrow S_E$$

(ii) $C_6H_{13}-\langle\rangle-\langle\rangle-OC_6H_{13}$

4-n-Hexyl-4'-n-hexyloxybiphenyl (S_E formed from a naturally-formed, mosaic B phase).
$$I \rightarrow S_B \rightarrow S_E$$

(iii) $C_8H_{17}O-\langle\rangle-\langle\rangle-CO.OC_2H_5$

Ethyl 4'-n-octyloxybiphenyl-4-carboxylate (normal ABE material); the B phase is a liquid crystal (hexatic) B phase.
$$I \rightarrow S_A \rightarrow S_B \rightarrow S_E$$

(iv) $\langle\rangle-\langle\rangle-CH=N-\langle\rangle-CH=CH-CO.OC_{10}H_{21}$

n-Decyl 4-(4'-phenylbenzylideneamino)cinnamate (normal ABE material); the B phase is a crystal B phase.
$$I \rightarrow S_A \rightarrow S_B \rightarrow S_E$$

(v) $\langle\rangle-\langle\rangle-CH=N-\langle\rangle-CH=CH-CO.OCH_2-\langle\rangle$
CH_3

3-Methylbenzyl 4-(4'-phenylbenzylideneamino)cinnamate (E phase after an AB-Isotropic transition); the B phase is a crystal B phase.
$$I \rightarrow S_{AB} \rightarrow S_E$$

X-ray diffraction pattern

The X-ray diffraction pattern for a smectic E phase is naturally very dependent on sample condition (powder sample, aligned domains). To present a diffractogram and say that it is typical could therefore be misleading. Basically however, a powder sample will give an inner ring and a number of sharp outer rings consistent with the ordered nature of this crystal smectic phase.

Differential scanning calorimetry

DSC and DTA indicate that transitions from the isotropic liquid, the S_A, or the S_B phases to the S_E phase are first order in nature. Enthalpy values are usually about 1–2 kcal mol^{-1} (4–8 kJ mol^{-1}).

References

Coates, D., Gray, G.W., and Harrison, K.J. (1973). *Mol. Cryst. Liq. Cryst.* **22**, 99.
Demus, D., and Sackmann, H. (1973). *Mol. Cryst. Liq. Cryst.* **21**, 239.
Diele, S., Brand, P., and Sackmann, H. (1972). *Mol. Cryst. Liq. Cryst.* **17**, 163.
Doucet, J., Levelut, A.-M., Lambert, M., Liebert, L., and Strzelecki, L. (1975). *J. Phys. (Paris)* **36**, 13.
Doucet, J. (1979). 'X-ray studies of ordered smectic phases.' In G.R. Luckhurst and G.W. Gray (eds.), *The Molecular Physics of Liquid Crystals*, Academic Press, London and New York, Chap. 14, pp. 317–341.
Goodby, J.W., and Gray, G.W. (1979). *Mol. Cryst. Liq. Cryst. Lett.* **49**, 165.
Gray, G.W., and Harrison, K.J. (1971a). *Symposium of the Chemical Society, Faraday Division* **5**, 54.
Gray, G.W., and Harrison, K.J. (1971b). *Mol. Cryst. Liq. Cryst.* **13**, 37.
Kölz, K.-H. (1966). Diplomarbeit Halle(S).
Leadbetter, A.J., Richardson, R.M., and Carlile, C.J. (1976). *J. Phys. (Paris)* **37**, 65.
Leadbetter, A.J., Frost, J.C., Gaughan, J.P., and Mazid, M.A. (1979a). *J. Phys. (Paris)* **40**, 185.
Leadbetter, A.J., Frost, J.C., and Mazid, M.A. (1979b). *J. Phys. (Paris)* **40**, 325.
Leadbetter, A.J., Mazid, M.A., and Malik, K.M.A. (1980). *Mol. Cryst. Liq. Cryst.* **61**, 39.
Levelut, A.-M., Doucet, J., and Lambert, M. (1974). *J. Phys. (Paris)* **35**, 773.
Richardson, R.M., Leadbetter, A.J., and Frost, J.C. (1978). *Ann. Phys.* **3**, 177.

6 The smectic F phase

Introduction

Until fairly recently, the smectic F phase was the rarest and most unusual of the various smectic polymorphic modifications. It was first discovered, together with the smectic G phase, by Demus and co-workers (1971), in just the one material

$$C_5H_{11}O-\underset{=N}{\overset{N=}{}}-C_5H_{11}$$

2-(4'-n-pentylphenyl)-5-(4''-n-pentyloxyphenyl)pyrimidine. Until 1978, this was in fact the only source of the exotic S_F phase, and because the synthesis of this compound is lengthy and supplies were limited, the structure and properties of the F phase remained relatively uninvestigated; something of a mystique therefore surrounded this modification for quite some time.

However, in 1978, detailed reports were made of new and more easily prepared materials which also exhibited the S_F phase. Goodby, Gray, and Mosley (1978) showed that some of the higher homologues of the terephthalylidene-bis-4-n-alkylanilines (the TBBA series) exhibited S_F phases, e.g. terephthalylidene-bis-4-n-pentylaniline (TBPA). In the same year, Dubois (1978) and Tinh and co-workers (1978) reported that a number of stilbenes also exhibited S_F phases; soon after, Goodby and Gray (1979) observed the S_F phase in the nO.m series for the compound N-(4-n-nonyloxybenzylidene)-4'-n-butylaniline (9O.4).

This variety of new materials exhibiting the S_F modification, particularly the readily available compound TBPA, provided the long awaited opportunity for more detailed physical studies of the phase to be made.

Several detailed structural studies of the S_F phase were indeed carried out shortly after these events, and these have led to a fairly sound understanding of the structure of the phase. As will emerge in a later chapter, these studies have also been instrumental in elucidating the structure of a further smectic modification—the smectic I phase—which for a time was confused with and sometimes mistakenly classified as the S_F phase or the tilted B phase. As we shall see, the structural difference between these uncorrelated smectic phases, each with a pseudo-hexagonal packing of the molecules within the layers, is essentially only one of tilt direction with respect to the hexagonal net.

Structure of the smectic F phase

The first structural study of the S_F phase was made by Demus and co-workers (1971) who discovered the phase in the compound 2-(4'-n-pentylphenyl)-5-(4"-n-pentyloxyphenyl)pyrimidine. The approach of Demus and his co-workers was, however, directed mainly towards proving the novelty and identity of the phases exhibited by this compound, because the S_F phase was indeed a new phase. Nonetheless, they were able to show that the ordering of the molecules within the layers of the S_F phase was of a tilted nature.

More detailed features of the structure of the S_F phase were first established by Leadbetter and co-workers (1979). Their studies involved TBPA

$$C_5H_{11}-\text{Ph}-N=CH-\text{Ph}-CH=N-\text{Ph}-C_5H_{11}$$

one of the new range of S_F materials. At the same time, they examined the racemic ester 80SI

$$C_8H_{17}O-\text{Ph}-\text{Ph}-CO.O-\text{Ph}-CH_2CH(CH_3)CH_2CH_3$$

at one time considered to be an S_F material, but later shown to be S_I in nature.

From their X-ray diffraction studies on well oriented samples of TBPA, it was possible to define the structure of the F phase as follows. The molecules are packed in layers with their long axes tilted with respect to the layer planes. For TBPA, the tilt angle mid-way in the temperature range (140–149°) of the S_F phase was about 23° (estimated from experimental measurements of the lamellar spacing). The tilt angle decreased gently with increasing temperature, and was greater than that in the higher temperature S_C phase which succeeds the S_F phase. The packing of the molecules in the layers was hexagonal. The diffraction maxima at high angle for the S_F phase were in the form of short bars; their width suggested a correlation length within the layers of 20–30 molecules (100–150 Å), and their length, a correlation length for the planes normal to the layers of about one layer. There is therefore poor correlation between the layers in

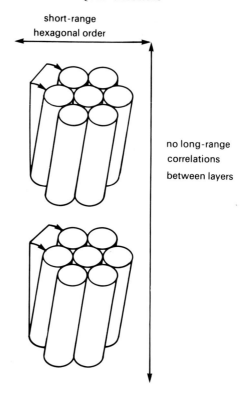

Figure 6.1 Structure of the S_F phase.

the sense that shift distortions of the hexagonal net occur between layers. Hence the hexagonal nets of the layers are out of register with one another. However, since mono-domains can be obtained with a uniform tilt direction, the hexagonal symmetry is apparently preserved through the bulk of the sample, and it may be concluded that the layers are free to slide over, but not to rotate relative to one another, i.e. the phase has extensive three-dimensional bond orientational ordering.*

This relatively fluid phase could therefore be regarded (Fig. 6.1) as a C-centred monoclinic cell having a regular hexagonal packing of the molecules in the layers. The detailed ordering was however restricted to about 30 molecular diameters within a layer and to 1 or 2 layers; the gross hexagonal symmetry was however preserved throughout the bulk sample.

---

*Bond orientational order is an expression for the register of the hexagonal net, i.e. the net is arranged in the same way from layer to layer but there are no real positional correlations of the molecules between layers.

THE SMECTIC F PHASE

Leadbetter and co-workers (1979) showed at the same time that one of the smectic phases of the ester 8OSI had a very similar structure to that of the S_F phase of TBPA, but that the tilt direction of the hexagonal net was different. For the S_F phase of TBPA this tilt was towards an edge of the hexagonal net, while for 8OSI it was towards a corner of the net. Later it emerged that the relevant phase for 8OSI was the S_I phase*—see Fig. 6.2.

These essential features of the S_F structure have since been confirmed and amplified by subsequent studies on a number of compounds (Benattar and co-workers, 1979; Leadbetter, Mazid, and Richardson, 1980; Goodby and co-workers, 1980a, b).

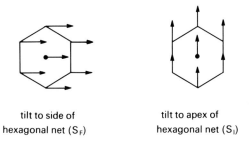

tilt to side of
hexagonal net (S_F)

tilt to apex of
hexagonal net (S_I)

Figure 6.2 Different tilt directions of the hexagonal net in S_F and S_I phases.

Benattar and co-workers (1979) again studied TBPA, carrying out their X-ray diffraction studies on mono-domain samples obtained by melting single crystals of the compound. They confirmed that the layers are almost uncorrelated, and that within the layers the order is pseudo-hexagonal, but much less well defined than that in the smectic phase which succeeds the S_F phase on cooling (i.e. in the S_G phase—see later). Studies of the Debye–Scherrer patterns gave less information about the S_F phase than about the S_G phase, but did show that the cell dimensions were roughly the same as those for the S_G phase (called at that time S_H by the authors). Only on this point did they disagree with Leadbetter and co-workers (1979) who had reported accurate cell parameters for the S_F phase. Because the ring was diffuse, Benattar and co-workers could only deduce the layer thickness (d) with accuracy. They reported d at 150° for the S_F phase to be 29.4±0.2 Å and calculated a tilt angle θ of 20±1°. For the S_G phase at 132°, they obtained $\theta=23.6\pm1°$.

Their studies also involved powder patterns for the S_F phase—two sharp reflections at 001 and 002 at small angles, and a broad diffuse ring at large

---

*Note that the abbreviation used for the ester 8OSI was 8OSF when the relevant smectic phase was thought to be S_F.

angles. The theoretical intensity profile was calculated and compared with that obtained by experiment, on the assumption that complete disorder exists between the layers and there is a perfect long-range order within the layers. The poorness of the fit showed that the disorder causing the broadening of the ring at large angles is not caused merely by lack of inter-layer correlation; loss of long-range order within the layers or lattice dynamic effects must also contribute to the situation, but the authors could not distinguish between these possibilities. They concluded that the S_F phase involves a *'two-dimensional structure'* with regard to positional order of the molecules, i.e. there is nearly complete absence of positional correlation between successive layers, while orientational correlation is maintained.

In a somewhat later paper by Benattar, Moussa, and Lambert (1980), further analysis of the intensity profiles obtained from powder samples was given, and it was concluded that the correlation length within the layers was about 250 Å, assuming the layers to be completely uncoupled.

There is therefore general agreement about the essential structure of the S_F phase; only more detailed points remain in any doubt. For example, the extent of the correlation within a layer cannot be quoted exactly ($\gtrsim 50$ Å) but, as pointed out by Gane, Leadbetter, and Wrighton (1981), this value will depend on the sample. In this paper by Gane and co-workers, the situation for the S_F phase is summarized in the following terms: the layers have long-range bond orientational order, but only short-range order (SRO) for the molecular positions within the layers. Both Gane and co-workers (1980) and Benattar and co-workers (1980) point out that this type of phase was predicted by Birgeneau and Litster (1978) for the S_B phase and that Halperin and Nelson (1978, 1979) predicted a tilted hexatic phase, to which the S_F and S_I phases in fact correspond. Halperin and Nelson in fact based their theory of two-dimensional melting on ideas of Kosterlitz and Thouless (1973) and predicted that melting could occur in two steps involving the loss successively of positional and orientational order.

Gane, Leadbetter, and Wrighton (1980) further confirmed the different tilt situations relating to S_F and S_I phases, by reporting on their studies of TBDA—the decyl homologue of TBPA—which exhibits both phase types:

$$C_{10}H_{21}-\bigcirc-N=CH-\bigcirc-CH=N-\bigcirc-C_{10}H_{21}$$

$$C \xrightarrow{73°} S_G \xrightarrow{120°} S_F \xrightarrow{150°} S_I \xrightarrow{156.1°} S_C \xrightarrow{190.8°} S_A \xrightarrow{192.4°} I$$

The only change which occurs at the transition from S_F (tilt direction towards an edge of the hexagon) to S_I (tilt direction towards an apex of the hexagon) is one of tilt direction. Both phases consist of uncorrelated layers, but with long-range bond orientational order. Both phases involve a tilted molecular arrangement in a pseudo-hexagonal molecular packing, there being only limited positional order within the layers. Therefore, the S_F phase is C-centred monoclinic with $a>b$, and the S_I phase is C-centred monoclinic with $a<b$. However, in a later study, Benattar and co-workers (1981) also found that the in-plane correlation length in the I phase was far greater than that in the F phase of TBDA.

Further detailed aspects of the structure and dynamics of the S_F phase must await neutron scattering studies.

The chiral smectic F phase

When a material that is optically active exhibits a S_F phase, then the phase itself is also optically active and is therefore called a chiral smectic F phase. The detailed structure of the chiral S_F phase has not as yet been studied, but presumably it will have a structure similar to that for the chiral S_C phase, but with a pseudo-hexagonal arrangement of the molecules in the layers. The tilts of the molecules would then be assumed to lie in the same direction in any one layer (for a particular domain), but on moving from one layer to the next, the tilt direction will be turned through a small angle in either a clockwise or an anti-clockwise direction. Thus, a helix will be formed by the tilt directions on passing through a succession of layers. From what has been said earlier about the relation between S_F and S_I phases, it will be understood that a chiral S_I phase can also exist, as discussed in a later chapter.

Textures of the smectic F phase

Both the S_F and the S_I phase show very similar microscopic textures, and when observed *in isolation* are very difficult to distinguish one from the other. The occurrence of both S_F and S_I phases in terephthalyl-idene-bis-4-n-decylaniline has however made it possible for direct comparisons of the minor differences between the textures of the phases to be drawn. This will be done in the chapter on the S_I phase. Such observations have only been made fairly recently, and it must be remembered that some of the earlier reports of S_F phases are now known to relate to S_I phases, simply because the two phases look very much alike. The two phases are also difficult to distinguish by miscibility methods.

Moreover, until recently, the S_F phase was found to be formed only on cooling (or heating) other phases, and therefore all of its textures were of a paramorphotic nature, i.e. determined by the texture of the preceding phase. Demus and Richter (private communication) have however shown that direct S_F to isotropic liquid transitions can occur; for example, in terephthalylidene-bis-4-n-butyloxymethoxyaniline, the formation of the S_F phase on cooling the isotropic liquid gives some indication of the natural textures of the S_F phase (see also p. 151).

The natural textures of the smectic F phase

When the smectic F phase separates directly from the isotropic liquid, it can exhibit two types of texture—the mosaic texture and a texture which may be of the cylindrical, spherulitic, or fan type.

The mosaic texture is not a true mosaic, but shows mosaic platelets separated by very fine lines (or optical discontinuities). The platelets have a tendency to contain *schlieren*-like brushes, but these do not appear to show actual intersections in the form of crosses. Thus, the texture is really of a mosaic-*schlieren* nature.

The cylindrical, spherulitic, or fan texture appears to consist of 'fans' viewed in the general direction of the hyperbola. These are without breaks or patches, and the texture has a disc-like appearance reminiscent of the polygonal texture. The discs each carry a familiar cross of extinction, but no radial lines run from the centre of the discs as in a normal fan texture. Hence it is unsure whether the disc is the top or bottom section of a cylinder, the smooth surface of a spherulite, or the smooth surface of a focal-conic domain with an almost circular cross-section (or an almost circular cross-section of such a domain). However, very few S_F to isotropic liquid transitions are known at present for pure compounds, and therefore we cannot make any generalizations about the natural texture of the S_F phase.

The paramorphotic textures of the smectic F phase

The smectic F phase can be formed on cooling a variety of other types of phase, and it can therefore exhibit a large number of paramorphotic textures. The S_F phase is most commonly obtained by cooling the smectic C phase. Thus, it too exhibits both a focal-conic and a mosaic-*schlieren* texture.

The focal-conic fan texture obtained in this way is very characteristic of the phase. The fans in the C phase are usually sanded and broken, the breaks being ill-defined and having a shaded appearance. On cooling to a smectic F phase, the sanding disappears and the breaks become well-defined and transform into black patches. These patches often have the characteristic shape of an elongated 'L'. Plate 35 shows the broken focal-conic fan texture of the smectic F phase of terephthalylidene-bis-4-n-pentylaniline (TBPA). In this plate, the fans run in an elongated fashion across the photomicrograph, and the distinctive 'L'-shaped patches can be seen running up and down, i.e. parallel to the layer planes, across the fan backs.

The S_F phase, as far as we can tell at present, appears to exhibit a *schlieren*-mosaic type of texture formed on cooling the *schlieren* texture of the S_C phase or the homeotropic texture of the S_A phase. In this texture, there are definite mosaic areas with rather fine lines of optical discontinuity dividing them; they also show some *schlieren* brushes, but with no point singularities.

Plate 36 shows the *schlieren*-mosaic texture of the smectic F phase of terephthalylidene-bis-4-n-pentylaniline (TBPA). In this texture, the mosaic platelets appear slightly undulating, with fine lines of discontinuity separating one domain from another. As stated above, the mosaic platelets show the brushes of a *schlieren* type of texture, but without the actual crosses or intersections of normal *schlieren* textures. The texture is therefore intermediate between the commoner types of mosaic and *schlieren* textures.

The smectic F phase can also be obtained directly from a smectic A phase—for example, with N-(4-n-nonyloxybenzylidene)-4'-n-butylaniline (9O.4). In this example of paramorphosis, the fan and mosaic textures are slightly different from those obtained by cooling a smectic C phase. The fan texture, although showing a similar broken-pattern across the fan backs, has a more 'chunky' appearance, as shown in Plate 37 for N-(4-n-nonyloxybenzylidene)-4'-n-butylaniline. The mosaic texture obtained for the phase from the homeotropic S_A is more like a true mosaic and does not show the *schlieren* features exhibited by the S_F phase formed from the S_C phase on cooling. Plate 38 shows the pale grey-blue mosaic texture of the smectic F phase obtained in this way for 9O.4.

Furthermore, and most unusually, the smectic F phase can be formed by cooling from more ordered phases. For example, Goodby and co-workers (1980b) showed that for N-(4-n-pentyloxybenzylidene)-4'-n-hexylaniline (5O.6), the S_F phase occurred *between* three-dimensionally ordered smectic B and smectic G phases. Thus, there is an order ⟷ disorder ⟷ order sequence on heating and cooling.

The microscopic textures that are exhibited by the F phase in these circumstances are very different from those normally observed. Firstly, the mosaic type of texture that is formed has very small mosaic areas. These tend to form in strings and give a hexagonal net-work before they fill in all the available space to make a complete mosaic texture. Plate 39 shows the very fine mosaic texture of the smectic F phase of 5O.6.

Secondly, the fan texture does not show the same number of breakages or 'L'-shaped patches as in the case of smectic F phases formed in more normal phase sequences. The fans appear lined (similarly to other smectic F phases), but only faintly, and they have a shattered appearance along their edges. Their appearance is also more mottled. Plates 76, 77, and 78 in Sequence 4 show the change in appearance of the fans in the B, F, G sequence for 5O.6.

The texture of the chiral smectic F phase

The smectic F phase is exhibited by suitable materials which are themselves optically active. Thus, the phase exhibited is said to be chiral. The textures of the chiral smectic F phase are of a more specialized nature.

The focal-conic fan texture obtained by cooling from a chiral smectic C phase is more broken than that of the achiral smectic F phase. Also, the breaks or 'L'-shaped patches are not simply black between crossed polarizers; they are highly coloured if the pitch of the phase is comparable with the wavelength of visible light. Thus, the backs of the fans have a speckled appearance. Plate 40 shows the focal-conic texture of the chiral smectic F phase of (+)-4-(2″-chlorobutanoyloxy)-4′-n-octyloxybiphenyl.

If the phase is formed by cooling a chiral smectic C phase in its pseudo-homeotropic texture (appearing black provided that the pitch is not comparable with the wavelength of visible light), then the chiral smectic F phase that forms will also show a pseudo-homeotropic texture. In this texture, the layers of molecules lie parallel to the glass supports, and the direction of the helical tilt director is vertical; this arrangement produces apparent homeotropy, i.e. the helices sit like coils on the glass surfaces. This texture does show some distinguishing marks in that it shows a kind of bubble texture. This texture shows areas or discs (maybe spherulites) bordered by optical discontinuities. These droplet areas appear to be in constant motion as the temperature is reduced. Moreover, the optical discontinuities appear to move as well, giving the impression of eddies rippling on the surface of water. This type of texture can be better observed in thicker preparations where it can appear pale blue or green in colour (Plate 41).

Identification and classification of the smectic F phase

Microscopic textures

(a) The smectic F phase was apparently easily recognized by its microscopic textures until the recent discovery of the smectic I phase which shows very similar textures. To be more certain of the identification of the F phase, careful miscibility studies have to be carried out with a *known* S_F compound. Difficulties can however arise, and the best method to distinguish the two phases is by X-ray diffraction.

(b) The S_F phase is probably most easily identified by its *schlieren*-mosaic texture. Although, both the S_I and S_F phases show similar textures of this type (which are certainly distinctive enough to categorize them from all other phases), the S_I and S_F textures do differ slightly and often sufficiently for the purposes of discrimination. These differences will be considered in more detail in the chapter on the S_I phase, but for the moment it may be noted that the S_F phase shows sharper lines of discontinuity at the edges of its platelet areas. Also, the S_I phase gives an ill-defined *schlieren* texture.

(c) The fan texture exhibited by the S_F phase is very different from those of the other tilted smectics which also form broken fan textures. The elongated 'L'-shaped patterns across the backs of the fans are very typical of the phase.

Miscibility studies

(a) Miscibility studies do pose some problems in the case of the smectic F phase. As already noted, the S_I and S_F phases are very similar in nature and texture, and they can also appear to be miscible. However, there always exist very, very narrow regions of immiscibility, although these can easily be overlooked. Thus, it is essential to perform miscibility studies using a wide range of binary mixtures covering small intervals of percentage change in composition. It has also been shown that it is possible to have an injection of smectic B properties into regions of smectic F phase in the diagram of state for binary mixtures involving S_I and S_F phases together. Furthermore, it is possible that the S_F phase can fall below the S_B phase in certain sequences, so giving a S_B–S_F–S_G sequence on cooling—as in the case of pure 5O.6 (see p. 101).

(b) Standard materials that exhibit the smectic F phase and are useful in miscibility studies.

(i) $C_5H_{11}O$—⌬—⌬—⌬—C_5H_{11}

2-(4'-n-Pentylphenyl)-5-(4"-n-pentyloxyphenyl)pyrimidine (the original F material).

$$I \rightarrow S_A \rightarrow S_C \rightarrow S_F \rightarrow S_G$$

(ii) C_5H_{11}—⌬—N=CH—⌬—CH=N—⌬—C_5H_{11}

Terephthalylidene-bis-4-n-pentylaniline (TBPA) (standard F material).

$$I \rightarrow N \rightarrow S_A \rightarrow S_C \rightarrow S_F \rightarrow S_G \rightarrow S_H$$

(iii) $C_9H_{19}O-\langle\bigcirc\rangle-CH=N-\langle\bigcirc\rangle-C_4H_9$

N-(4-n-Nonyloxybenzylidene)-4'-n-butylaniline (90.4) (A to F compound).

$$I \to S_A \to S_F \to S_G$$

Differential scanning calorimetry

DSC and DTA indicate that the S_C to S_F transition is second-order or weakly first-order in nature and that the S_F to S_G transition is first-order. The approximate value for the enthalpy for the S_C to S_F transition is between 0.3 and 0.6 kcal mol^{-1} (1.2 to 2.4 kJ mol^{-1}).

X-ray diffraction pattern

The reader is directed to p. 129 where the X-ray diffraction of smectic F and of smectic I phases is conveniently discussed together.

References

Benattar, J.J., Doucet, J., Lambert, M., and Levelut, A.-M. (1979). *Phys. Rev. A* **20**, 2505.
Benattar, J.J., Moussa, F., and Lambert, M. (1980). *J. Phys. (Paris)* **41**, 1371.
Benattar, J.J., Moussa, F., and Lambert, M. (1981). *J. Phys. (Paris) Lett.* **42**, 67.
Birgeneau, R.J., and Litster, J.D. (1978). *J. Phys. (Paris) Lett.* **39**, 399.
Demus, D., Diele, S., Klapperstück, M., Link, V., and Zaschke, H. (1971). *Mol. Cryst. Liq. Cryst.* **15**, 161.
Dubois, J.C. (1978). *Ann. Phys.* **3**, 131.
Gane, P.A.C., Leadbetter, A.J., and Wrighton, P.G. (1981). *Mol. Cryst. Liq. Cryst.* **66**, 247.
Goodby, J.W., Gray, G.W., and Mosley, A. (1978). *Mol. Cryst. Liq. Cryst.* **41**, 183.
Goodby, J.W., and Gray, G.W. (1979). *Mol. Cryst. Liq. Cryst. Lett.* **56**, 43.
Goodby, J.W., Gray, G.W., Leadbetter, A.J., and Mazid, M.A. (1980a). *J. Phys. (Paris)* **41**, 591.
Goodby, J.W., Gray, G.W., Leadbetter, A.J., and Mazid, M.A. (1980b). 'The smectic phases of the N-(4-n-alkoxybenzylidene)-4'-alkylanilines (nO.m's)—some problems of phase identification and structure.' In W. Helfrich and G. Heppke (eds.), *Liquid Crystals of One- and Two-Dimensional Order, Springer Series in Chemical Physics 11*, Springer-Verlag, Berlin, Heidelberg, and New York, pp. 3–18.
Halperin, B.I., and Nelson, D.R. (1978). *Phys. Rev. Lett.* **41**, 121.
Halperin, B.I., and Nelson, D.R. (1979). *Phys. Rev. B* **19**, 2457.
Kosterlitz, J.M., and Thouless, D.J. (1973). *J. Phys. C.* **6**, 1181.
Leadbetter, A.J., Gaughan, J.P., Kelly, B., Gray, G.W., and Goodby, J.W. (1979). *J. Phys. (Paris)* **40**, 178.
Leadbetter, A.J., Mazid, M.A., and Richardson, R.M. (1980). 'Structure of the smectic B, F, and H phases of the N-(4-n-alkoxybenzylidene)-4'-n-alkylanilines and the transitions between them.' In S. Chandrasekhar (eds.), *Liquid Crystals*, Heyden, London, Philadelphia, and Rheine, pp. 65–79
Tinh, N.H., Zann, A., Dubois, J.C., and Billard, J. (1978). *J. Phys. (Paris)* **39**, 1283.

7 The smectic G phase

Introduction

The smectic G phase was discovered by Demus and co-workers (1971) in the compound 2-(4'-n-pentylphenyl)-5-(4"-n-pentyloxyphenyl)pyrimidine

$$n\text{-}C_5H_{11}O-\!\!\bigcirc\!\!-\!\!\bigcirc\!\!\substack{N\\=N}\!\!-\!\!\bigcirc\!\!-C_5H_{11}\text{-}n$$

which was also the source of the first known S_F phase. However, for the reasons outlined below, certain problems over the nomenclature of the S_G phase then arose.

Shortly after the discovery of the S_G phase, de Vries and Fishel (1972) published the results of studies by X-ray diffraction of the smectic phase of N-(4-n-butyloxybenzylidene)-4'-ethylaniline (4O.2), a member of the nO.m series. They concluded that this phase which possessed a three-dimensional, tilted, herringbone structure was of a novel kind, and suggested that it be given a new code letter, S_H. However, later studies were to show that this phase was co-miscible with the S_G phase of the pyrimidine.

The situation was further complicated by the spread of the mistaken belief that the S_H phase of 4O.2 and other compounds in which the phase was ultimately found were co-miscible with orthogonal S_B phases. This arose through the failure to observe narrow regions of immiscibility which separate the two phases. As a result, these phases became known as tilted S_B phases and, for obvious reasons, they were frequently coded as S_{B_C} phases. It therefore became widely accepted that

$$S_B \text{ tilted} \equiv S_{B_C} \equiv S_H$$

but the use of the code letter H was not general, simply because the phases were believed to be in the same miscibility class as the orthogonal S_B phase.

At that time, the S_G phase of the pyrimidine—generally thought to be different from the S_H phase—became identified (Richter, Demus, and Sackmann, 1976) with the tilted S_E phase instead of with the tilted S_B phase. As the pyrimidine compound is more difficult to prepare than say 4O.2, the usage of the label $S_H(S_{B_C})$ spread, because miscibility studies involving the pyrimidine were not commonplace due to sample scarcity.

Fortunately this confusing situation persisted for only about three years, when Doucet and Levelut (1977) and de Jeu and de Poorter (1977) showed that reversible transitions between orthogonal S_B and tilted S_B phases can occur in pure compounds. At about this time Goodby and Gray were examining more closely the miscibility relationship between orthogonal and tilted S_B phases, and two years later Goodby and Gray (1979) published results demonstrating clearly that these phases are in fact immiscible, a conclusion substantiated by Sackmann (1979).

It seemed wrong therefore that the tilted phase should be associated with the code letter B. Because of the prevailing view that the S_G phase was related to a tilted E phase, and consequent upon the earlier work of de Vries and Fishel (1972) on 4O.2, Goodby and Gray (1979) elected to recommend the use of S_H to code those phases hitherto called tilted S_B or S_{B_C}. However, Richter and co-workers (1976) had shown that the S_G phase of the pyrimidine and the smectic phase of 4O.2 were *miscible*. As the Halle group had discovered this phase in the pyrimidine, they naturally described 4O.2 as

$$I \to N \to S_G$$

As the discovery of the S_G phase (1971) predates the use of the code letter H for the smectic phase of 4O.2 (1972), it has been agreed (Demus, Goodby, Gray, and Sackmann, 1980) that the tilted S_B ($\equiv S_{B_C} \equiv S_H$) phase should henceforth be called S_G. Acceptance of this has seen the end of a period in which many counterproductive misunderstandings had arisen through the use of different code letters (G/H) by different research groups to refer to one and the same type of smectic phase, e.g. in 4O.2 and other nO.m's. As pointed out by Goodby, Gray, Leadbetter, and Mazid (1980), this duality of nomenclature had however even more serious consequences, because as a secondary outcome, other workers had come to use the code letter G to refer to the tilted S_E phase, which the Halle group subsequently referred to as S_H. In some cases therefore a total inversion of nomenclature between H and G arose, whereas in other cases, the two code letters were being used synonymously.

The unified nomenclature scheme was laid down in discussions with Sackmann and Demus during a visit by the authors (Goodby and Gray) to Halle, and cemented in place at the ensuing Conference on Liquid Crystals of One- and Two-Dimensional Order held in Garmisch-Partenkirchen in January 1980.

Acceptance that the letter G be used for the phase originally described by many as a tilted B phase, and that H be used for the phase that is a tilted analogue of an E phase, has led to the general use of the coding sequences below to describe the phase behaviours of three very important smectogens.

TBBA	I N S_A S_C S_G S_H
Pyrimidine	I S_A S_C S_F S_G
4O.2	I N S_G

Structure of the smectic G phase

From detailed X-ray diffraction studies by Doucet and Levelut (1977), Levelut, Doucet, and Lambert (1974), Levelut (1976), and Leadbetter and co-workers (1979), it has been shown that the molecules in the G phase are packed in layers. The molecules within the layers have their long axes tilted with respect to the normal to the layer planes; the average tilt angle of the molecules is approximately 25–30°, and this angle usually increases very slowly with decreasing temperature.

The S_G phase has a C-centred monoclinic cell, the molecules having a pseudo-hexagonal close packing; in the plane at right angles to the tilt direction, this close packing arrangement is of the hexagonal kind. The layers are also very well correlated giving a three-dimensional structure. Although it is known (Doucet, 1979; Gane and co-workers, 1981) that considerable dynamic disorder of the molecules exists in the S_G phase, the inter-layer correlations are well established, since all S_G phases give a number of *hkl* reflections. The inter-layer correlations for all S_G phases so far studied appear to involve an AAA--- monolayer stacking.

Thus, the phase can be considered as a crystal phase, and in support of this, the S_G phase does not show any phonons associated with layer shearing or undulation modes. Therefore, the layers do not show a rippling, undulating motion as they do in some related B phases. Indeed, S_B to S_G transitions in the nO.ms (Goodby and co-workers, 1980) are a consequence of the undulating motion in the S_B phase reaching a critical amplitude when the longitudinal displacements of the molecules become locked into the tilted S_G structure.

Apparently, therefore, the phase is of a more crystalline nature than the correlated S_B phase. Figure 7.1 shows the structure within the layers of the S_G phase, and Fig. 7.2 the layered structure of the S_G phase.

The dynamics of the molecular motion in the S_G phase have not been examined fully, but it is usually assumed that completely free rotation of the molecules is impossible. This is argued on the grounds that the molecular packing in the pseudo-hexagonal net is too close to allow unrestricted motion about the molecular long axes. However, neutron

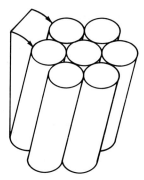

Figure 7.1 Hexagonal close-packed molecules in the smectic G phase (tilt direction towards a side of the pseudo-hexagonal net).

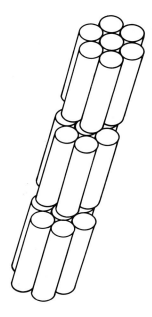

Figure 7.2 The correlated three-dimensional structure of the smectic G phase. The direction of the tilt should be seen as towards a side of the repeating pseudo-hexagonal net (see Fig. 7.1).

scattering studies by Leadbetter, Mazid, and Richardson (1980) show that the molecules undergo rapid reorientational motion about their long axes on a time scale of 10^{-11} s. Normally, therefore, the molecules are envisaged as having co-operative or oscillatory movement about their long axes. The local structure is a herringbone or chevron packing, and the average hexagonal symmetry is obtained by a superposition of three different orientations of the local orthorhombic cells. Thus the situation is a dynamic one, individual molecules reorientating rapidly about six equivalent sites on the time scale given above. The position is similar to that already discussed for the S_B phase, and at any given moment in time the molecules will have a chevron-like packing.

In the case of the S_G phases of the nO.ms, neutron scattering studies have indicated a diffusive motion of the molecules in the direction of the molecular long axis—$<Z^2>^{\frac{1}{2}}=1.09$ Å—on a time scale similar to that for the rotational motion. This suggests that the two motions are coupled.

Although the S_I phase has not yet been discussed in any detail, it was necessary in the chapter on the S_F phase to draw attention to the fact that these two phases are closely related. Both phases have C-centred monoclinic cells with a pseudo-hexagonal packing of the molecules whose long axes are tilted with respect to the normal to the layer planes; moreover, the layers of both phases are almost uncorrelated. The two phases differ only in the tilt direction with respect to the hexagonal net. For S_F, the tilt is towards an edge of the hexagon and for S_I, the tilt is towards an apex of the hexagon.

From what has been said up till now, it will be realized that the S_G phase is a correlated analogue of either a S_F or a S_I phase, and this immediately gives rise to the question as to whether two different S_G phases are possible, one with a tilt to the edge of the hexagon (Fig. 7.3(a)) and one with a tilt to the apex of the hexagon (Fig. 7.3(b)). There is no doubt in fact that this is so. When the S_G phase is formed by cooling a S_F phase, the tilt direction with respect to the hexagonal net is retained (tilt towards the edge of the hexagon as in Fig. 7.3(a)). This occurs with terephthalyl-idene-bis-4-n-pentylaniline (TBPA).

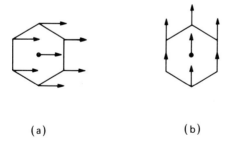

(a)   (b)

Figure 7.3 Tilt directions with respect to the hexagonal net for (a) S_G, (b) $S_{G'}$.

However, the G analogue formed on cooling a S_I phase also retains the tilt direction of the precursor phase (tilt towards an apex of the hexagon as in Fig. 7.3(b)). Therefore the G analogue formed on cooling the S_I phase of racemic 4-(2'-methylbutyl)phenyl 4'-n-octyloxybiphenyl-4-carboxylate (8OSI) has the opposite tilt to that of the S_G phase of TBPA. As suggested by Gane and co-workers (1981), we here refer to the G phase of 8OSI as a G' phase, since at that time it had not been proved* that the G and G' modifications were immiscible. Therefore, although the structural feature distinguishing S_G and $S_{G'}$ phases is the same as that which distinguishes S_F and S_I phases, different code letters seemed justified only in the latter case where the F and I phases were immiscible, and indeed examples were known in which reversible S_F to S_I transitions occur in pure compounds.

The literature of course contains very many examples of G-type smectic phases—all the phases originally described as tilted S_B are G in type. Moreover, such phases are formed on cooling many different phases—not only S_F and S_I phases. Therefore a considerable amount of work needs to be done before it is known just how many of these phases are S_G and how many are $S_{G'}$ in nature.

At this point it is interesting to stress that the layer correlated analogue of S_F is S_G and that of S_I is $S_{G'}$, and that the occurrence (or breakdown) of the layer correlation in each case marks a true phase transition between two immiscible smectic types ($S_F \leftrightarrow S_G$ and $S_I \leftrightarrow S_{G'}$). Now the presence or absence of layer correlation is *again* the distinguishing feature between correlated, crystal S_B phases and uncorrelated, liquid crystal S_B phases. At the time of writing, the two B types had not been shown to be immiscible, and so the code letter B is still retained for both—but see Chapter 10 for recent developments.

Structure of the chiral smectic G phase

If the constituent molecules of a material which exhibits a smectic G phase are of a chiral nature, then the phase itself may also be weakly optically active; it is then termed a chiral G phase. Coates and Gray (1976) showed that (+)-terephthalylidene-bis-(4'-methylhexyloxy)aniline (TBMHA) exhibited a chiral smectic B phase. The nomenclature system at that time named all tilted B (G) and orthogonal B phases by the same letter B. The viscous B phase persistently formed a mosaic texture from which it was not possible to obtain conclusive proof of the optical activity of the phase. However, the B phase formed by mixtures of TBMHA with an achiral diluent (capable of giving either a tilted smectic B or an orthogonal smectic B) were shown to exhibit optically active properties. Extrapolation of this situation indicated that pure TBMHA must also be capable of forming an optically active B phase. For the B phase to be chiral the molecules must be tilted with respect to the layer planes. Therefore the B phase of TBMHA is of the tilted B type. Consequently, Coates and Gray actually showed that it

---

*Eventually (see p. 140) it was shown that the G and G' modifications are immiscible, and it became necessary to introduce a new code letter, S_J, for $S_{G'}$.

was possible for a smectic G phase to be chiral in mixtures (see also Chapter 10).

As pointed out by Gane and co-workers (1981), the phase described as smectic IV by Doucet and co-workers (1978) for the optically active compound

$$C_6H_{13}O-\underset{}{\bigcirc}-CH=N-\underset{}{\bigcirc}-CH=CH-CO.OCH_2CH(CH_3)Cl$$

$$I \to S_A \to S_{C^*} \to S_{III} \to S_{IV}$$

must also be of the G type. However, only if S_{III} were S_{F^*}, could S_{IV} be S_{G^*}. The results of Gane and co-workers, subsequently confirmed by Pindak, 1982, reveal that S_{III} is in fact S_{I^*}, and so S_{IV} should be $S_{G'^*}$—but see below and Chapter 10.

Few structural studies had been carried out at that time on chiral smectic G phases, and it was originally simply presumed that the structure of the phase is similar to that of chiral C, I, and F phases. In this case, the molecules would be hexagonally close packed in layers within each of which the tilts must be in the same direction. In the layers immediately above and below an object layer, this tilt direction will however be turned through a small angle. Thus, on passing from layer to layer, the tilt direction will turn slowly either in an anti-clockwise or a clockwise direction, dependent upon the sign of the optical asymmetry of the system. This would give a helical change in tilt direction, as described earlier for the S_C and S_F phases.

However, it was appreciated that the correlations between the layers in this modification along with the pitch of the helix will determine whether the chiral G phase does occur or not. If the correlations are extremely strong, then it will be more likely that the pitch length will be extremely large, and therefore, detection of the optical rotatory properties of the phase will be difficult. For a weak layer correlation coupled with a stronger twisting power of the molecules, then it will be more likely that a G^* phase will be formed. However, if the correlations are extremely weak, one might envisage formation of a structure similar to that of the chiral F. In this way it was possible to see how the structure of the phase may modify the optically active character of the chiral phase. For example, in a system where we have S_{C^*}, S_{F^*} and S_{G^*} phases, it is possible that they may become less optically active on passing through the sequence, S_{C^*} having a tighter helix than S_{F^*} which may have a tighter helix than S_{G^*}. However, as discussed further in Chapter 10, all single-component crystal G phases that have been studied by X-ray diffraction possess no helical arrangement of the tilt direction. Mixed G phases can however behave differently, as mentioned above.

Textures of the smectic G phase

The smectic G phase exhibits numerous microscopic textures because it is formed from a large variety of precursor phases on cooling. It is more usual to observe the formation of the phase on cooling a C phase, but it can be formed from nematic, S_A, S_B, and S_F phases as well.

As mentioned earlier, a G phase with an opposite tilt direction with respect to the hexagonal net, i.e. a G' phase, has recently been recognized and may be formed on cooling a S_I phase. The textural differences between G and G' phases await a full study, but some comments on this will be made inthe chapter on S_I phases where the ester 8OSI with an I↔G' transition will be discussed.

The natural texture of the smectic G phase

Very few examples of materials which exhibit a direct isotropic to smectic G phase transition are known. Thus the range of natural textures observed for the G phase is limited. On cooling from the isotropic liquid, the G phase nucleates *via* a dendritic growth pattern (Plate 42) and subsequently forms mosaic platelets (Plate 43).

However, there are numerous examples of formation of the G phase directly from the nematic phase, so providing other indications of its natural textures. The S_G phase formed on cooling the nematic phase separates in large 'splinter'-shaped particles or lancets. These are often oblong in shape, almost needle-like, and they coalesce to form a mosaic texture which consists of oblong platelets. Plate 44 shows the natural texture of the S_G phase of *N*-(4-n-butyloxybenzylidene)-4'-ethylaniline (4O.2) separating from the nematic phase on rapid cooling. It is interesting to note how the *schlieren* texture of the nematic phase shrinks and separates quickly away from the newly-formed phase. The fact that the phase separates from the nematic phase in the form of mosaic splinters is an indication that the phase is very ordered and almost of a crystalline nature.

The paramorphotic textures of the smectic G phase

The smectic G mesophase exhibits a number of paramorphotic smectic textures. These are however derived from basically two textures, namely the focal-conic fan and the homeotropic or *schlieren* textures. Firstly, the fan texture of the G phase can be obtained by cooling S_A, S_B, S_C, or S_F phases. In all of these cases, except for that involving the B phase, the fan textures obtained are rather similar. Secondly, the mosaic texture of the G phase shows a variety of forms depending on the precursor phase, except that this time, the mosaic textures formed when the phase is obtained from the B and the A phases are somewhat dissimilar to those formed from the other phases.

The focal-conic fan textures of the smectic G phase

There are basically two types of focal-conic fan textures shown by the G phase—the patchwork or broken fan and the arced-broken fan textures. The patchwork texture is normally obtained on cooling A, C, or F phases. The fan backs become broken at the transition to the S_G phase. The breakages take on the form of large dark patches, giving the texture a chequered pattern. These dark regions become lighter on rotation of the analyser and polarizer, indicating that they are not homeotropic, and are due to a particular orientation of the molecules. Sometimes, these areas can be quite bright and are marked by dark lines of optical discontinuity. The typical broken fan texture of the smectic G phase is shown in Plate 45 for terephthalylidene-bis-4-n-butylaniline (TBBA). In this photomicrograph, the S_G phase was obtained by cooling a S_C phase. It is also interesting to note that this texture, although similar to that of the S_F phase, can be used to distinguish quite easily between the two. Normally, the fan texture of the S_G modification shows rectangular breakages or patches and has a more chunky appearance, whilst the breakages for the S_F phase are smaller and are more irregular in shape, with the dark areas showing almost a striped pattern.

The broken, arced, focal-conic fan texture of the S_G phase is only observed on cooling the S_B phase, and this is characteristically shown by the N-(4-n-alkoxybenzylidene)-4'-n-alkylanilines (nO.ms). In this texture, the S_G phase forms from a three-dimensional B phase, and therefore, as the layers tilt, the back of the fan breaks into parallel arcs running in the same direction as the layers. Thus, the fan has a similar appearance to that of the arced S_E fan. However, in the case of the G fan, these arcs are not continuous from one fan area to another, and show breaks, probably due to the fact that the tilt direction may change within one of the bands running across the back of the fan. The arced, focal-conic fan texture of the smectic G mesophase of N-(4-n-heptyloxybenzylidene)-4'-n-pentylaniline (7O.5) is shown in Plate 75, Sequence 3.

The mosaic texture of the smectic G phase

The S_G phase shows a variety of mosaic textures depending on the paramorphosis of the system. In general, however, the mosaic textures have a similar type of structure which is closely related to that of the mosaic texture of the S_B phase (Fig. 2.12). It is probable, therefore, that the mosaic texture of the S_G modification consists of large individual platelet areas in each of which the molecules have a preferential tilt direction as shown in Fig. 7.4.

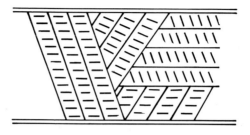

Figure 7.4 Probable arrangement in the mosaic texture of the smectic G phase.

The major differences in the mosaic textures formed by the G phase are connected with the related structure of the precursor phase. For example, if the mosaic texture of the S_G phase is obtained by cooling a S_B phase, then the S_B phase must be in its homeotropic texture. Therefore, the layers in the S_B phase are parallel to the glass support. In the case of the N-(4-n-alkoxybenzylidene)-4'-n-alkylanilines (nO.ms), it is known that S_G phase formation occurs *via* a tilting of the *layers*. Therefore whilst the molecules in the bulk of the phases are probably still orthogonal to the glass surfaces and the layers are tilted with respect to the surfaces (Fig. 7.5), the situation in the layers immediately adjacent to the glass will depend upon the strength of the surface forces. With weak interactions, the tilt of the layers may propagate through the film from one surface to the other. With strong interactions, the parallel distribution may persist in a few layers next to the glass, and in these layers the molecules may simply tilt over slightly (Fig. 7.5).

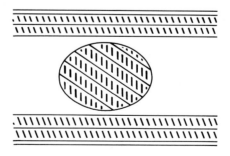

Figure 7.5 Possible model for the development of the mosaic texture of the S_G phase in the homeotropic texture of the S_B phase when strong surface forces are involved.

The mosaic areas are not particularly large in this form of texture. Also, they are not very highly coloured, presumably because the molecules in the bulk are not in a very tilted orientation. Hence, the mosaic texture shows small, ill-defined platelet areas that are not particularly birefringent. An example of this texture of the smectic G phase of N-(4-n-heptyloxybenzyl-idene)-4'-n-pentylaniline (7O.5) is shown in Plate 75 of Sequence 3.

When the S_G phase is formed from the S_A phase on cooling, the mosaic texture obtained shows similar features to those already discussed for the situation involving cooling from the S_B phase. However, the mosaic areas are a lot larger, possibly because the low viscosity of the S_A phase allows the molecules just to tilt, therefore, eliminating the effect of the layers tilting in the bulk.

Cooling from the S_C or S_F phases produces the mosaic texture which is more commonly associated with the S_G phase. The mosaic platelets are quite large, and they are highly coloured; therefore, the structure of the texture is probably similar to that shown in Fig. 7.4. There are a number of examples of this texture shown in the section containing the photographic plates, e.g. Plate 46 shows the mosaic texture of the S_G phase of terephthalylidene-bis-4-n-butylaniline (TBBA) and Plate 65 of Sequence 1 shows the mosaic texture of 4-(2'-methylbutyl)phenyl 4'-n-octyloxy-biphenyl-4-carboxylate (8OSI).

The textures of the chiral smectic G phase

Chiral S_G phases exhibit two types of microscopic textures; these are the broken focal-conic fan and the mosaic textures. In the broken focal-conic fan texture, the fan backs are broken in either a patchwork or more usually a banded fashion. It is interesting to compare the texture with those of the chiral S_C and S_F phases. In these two phases the fans are coloured and show almost a rainbow effect across their backs stemming from selective reflections from the helical structure of the phase. In the S_G phase, however, this colouring effect disappears, and the fans are mainly monotone. This contrast is shown for (+)-4-(2'-methylbutyl)phenyl 4'-n-octyloxybiphenyl-4-carboxylate (8OSI*) in Sequence 2 where Plate 68 shows the S_A fans, Plate 69 the S_C* fans, Plate 70 the S_I* fans, and Plate 71 the S_G* fans. This loss of colour may indicate that the three-dimensional structure of the S_G phase partly interferes with the formation of the helical structure. The pitch length therefore increases and the selective reflections of light that would normally give the colouring effect are no longer in the visible range. Hence the fans become monotone in colouration.

The mosaic texture of the chiral S_G phase is almost identical with those mosaic textures of the achiral S_G phase formed on cooling achiral S_C or S_F phases. The chiral S_G phase is normally formed by cooling S_{C^*} or S_{F^*}

phases which usually form some kind of pseudo-homeotropic texture, because of the helical nature of the structure. However, the S_G phase is somewhat limited in this respect by its own three-dimensional structure. This tends to unwind the helical structure and the S_{G*} phase adopts a more usual mosaic texture which gives the impression that the phase is almost achiral (Plate 47).

Identification and classification of the smectic G phase

Microscopic textures

(a) The smectic G phase usually exhibits one natural texture, the mosaic texture. This texture characteristically forms *via* dentritic growth of elongated platelets or splinters as the mesophase separates from the preceding medium. The resulting mosaic texture consists of elongated platelet areas that have an almost rectangular shape. The phase is very viscous, but it can be sheared. Mechanical displacement of the coverslip whilst the phase is in this texture shows that the mosaic areas do not break up, but tend to crumple into one another whilst retaining some of their original shape.

(b) The S_G phase shows numerous paramorphotic textures each of which has its own characteristics. When formed from a S_B phase on cooling, the S_G phase shows an arced, broken, focal-conic fan texture, and also a mosaic texture that has small ill-defined platelet areas that are weakly coloured. If however, the phase is obtained by cooling a S_C or a S_F phase then the fan texture is broken, but with a patchwork pattern, and has a chequerboard appearance. The mosaic texture is usually made up of large, rounded, platelet areas that are often highly coloured.

Miscibility studies

(a) Miscibility studies involving the smectic G phase have been the source of numerous problems in phase identification.

Firstly, the tilted smectic G phase and the orthogonal smectic B phase can often appear to be co-miscible. However, truly continuous regions of S_G and S_B phases are never in fact obtained and at some point in the miscibility diagram of state there will be a small region of immiscibility (~2–3% of the composition diagram). Detailed investigations have to be made in such cases so that this region is not missed. A typical texture which shows the immiscibility of the two phases is shown in Plate 48 for the S_G phase of terephthalylidene-bis-4-n-butylaniline (TBBA) and the S_B phase of 4-n-pentyloxyphenyl 4′-n-octyloxybiphenyl-4-carboxylate.

Secondly, two materials that exhibit S_G phases can sometimes show an injection of an orthogonal S_B phase in the 40 to 60% region of the phase diagram. Often this S_B region is bounded at its lower temperature limits by S_B to S_G transitions, making an island of B phase in the diagram of state. However, sometimes this lower boundary is not detected, and therefore one can progress from S_G to S_B to S_G again on moving across the diagram with changing concentration.

As pointed out earlier, the realization that two types of S_G phase can exist raised the question as to whether S_G and $S_{G'}$ phases are or are not miscible. At present, the assignment of S_G phases to the G or G' types has been made for relatively few materials and the experimental work required to establish the miscibility situation for S_G and $S_{G'}$ phases has yet to be done—but see Chapter 10.

(b) Standard materials that exhibit S_G phases and may be useful in miscibility studies are given below. In only one case, (v), is it known with certainty that the phase is $S_{G'}$.

(i) $\quad$ n-C$_4$H$_9$O—⟨◯⟩—CH=N—⟨◯⟩—C$_2$H$_5$

N-(4-n-Butyloxybenzylidene)-4'-ethylaniline (4O.2) (this material shows the G phase in its natural texture).

(ii) $\quad$ n-C$_4$H$_9$—⟨◯⟩—CH=N—⟨◯⟩—N=CH—⟨◯⟩—C$_4$H$_9$-n

Terephthalylidene-bis-4-n-butylaniline (TBBA) (the G phase occurs below a C phase at a high temperature; this material is a very well-known standard, although the nature of S_{VII} is unclear).

$$I \to N \to S_A \to S_C \to S_G \to S_H \to S_{VII}$$

(iii) $\quad$ n-C$_7$H$_{15}$O—⟨◯⟩—CH=N—⟨◯⟩—C$_5$H$_{11}$-n

N-(4-n-Heptyloxybenzylidene)-4'-n-pentylaniline (7O.5) (the G phase forms from a B phase on cooling).

$$I \to N \to S_A \to S_C \to S_B \to S_G$$

(iv) $\quad$ n-C$_5$H$_{11}$O—⟨◯⟩—⟨◯⟩—C$_5$H$_{11}$-n

2-(4'-n-Pentylphenyl)-5-(4''-n-pentyloxyphenyl)pyrimidine (PPOP) (the original smectic G material).

$$I \to S_A \to S_C \to S_F \to S_G$$

(v)   n-C$_8$H$_{17}$O—⟨⟩—⟨⟩—CO.O—⟨⟩—CH$_2$CH(CH$_3$)C$_2$H$_5$

4-(2'-Methylbutyl)phenyl 4'-n-octyloxybiphenyl-4-carboxylate (8OSI) (the G' phase occurs below a S$_I$ phase).

$$I \rightarrow N \rightarrow S_A \rightarrow S_C \rightarrow S_I \rightarrow S_G \rightarrow S_{H'}$$

X-ray diffraction pattern

The appearance of the X-ray diffraction pattern of the smectic G phase depends upon the quality of alignment of the phase. For simple unaligned samples (powder samples), the diffraction pattern is similar to that of a S$_B$ phase and consists of a sharp inner and a sharp outer ring. Additionally, however, weaker outer rings may be seen close to the strong ring. The layer spacing may be obtained from the position of the inner ring, and from this, and a knowledge of the molecular length, the tilt angle may be derived.

Differential scanning calorimetry and differential thermal analysis

Differential scanning calorimetry or differential thermal analysis shows that transitions to the S$_G$ phase are usually first order in type, with enthalpies of 0.5 to 0.75 kcal mol^{-1} (2–3 kJ mol^{-1}). However, it is interesting to note that the enthalpy for the S$_C$ to S$_F$ transition is much greater than that of the S$_F$ to S$_G$ transition which is usually a weak first order transition of enthalpy ~0.25 kcal mol^{-1} (~1 kJ mol^{-1}).

References

Coates, D., and Gray, G.W. (1976). *Mol. Cryst. Liq. Cryst. Lett.* **34**, 1.
De Jeu, W.H., and de Poorter, J.A. (1977). *Phys. Lett.* **61A**, 114.
Demus, D., Diele, S., Klapperstück, M., Link, V., and Zaschke, H. (1971). *Mol. Cryst. Liq. Cryst.* **15**, 161.
Demus, D., Goodby, J.W., Gray, G.W., and Sackmann, H. (1980). *Mol. Cryst. Liq. Cryst. Lett.* **56**, 311, and 'Recommendation for the use of the code letters G and H for smectic phases', in W. Helfrich and G. Heppke (eds.), *Liquid Crystals of One- and Two-Dimensional Order, Springer Series in Chemical Physics 11,* Springer-Verlag, Berlin, Heidelberg, and New York, pp. 31–33.
De Vries, A., and Fishel, D.L. (1972). *Mol. Cryst. Liq. Cryst.* **16**, 311.
Doucet, J., and Levelut, A.-M. (1977). *J. Phys. (Paris)* **38**, 1163.
Doucet, J., Keller, P., Levelut, A.-M., and Porquet, P. (1978). *J. Phys. (Paris)* **30**, 548.
Doucet, J. (1979). 'X-ray studies of ordered smectic phases.' In G.R. Luckhurst and G.W. Gray (eds.), *The Molecular Physics of Liquid Crystals,* Academic Press, London and New York, Chap. 14, pp. 317–341.
Gane, P.A.C., Leadbetter, A.J., and Wrighton, P.G. (1981). *Mol. Cryst. Liq. Cryst.* **66**, 247.
Goodby, J.W., and Gray, G.W. (1979). *J. Phys. (Paris)* **40**, 363.
Goodby, J.W., Gray, G.W., Leadbetter, A.J., and Mazid, M.A. (1980). 'The smectic phases of the N-(4-n-alkoxybenzylidene)-4'-n-alkylanilines (nO.ms)—some problems of phase identification and structure.' In W. Helfrich and G. Heppke (eds.), *Liquid Crystals of One- and Two-Dimensional Order, Springer Series in Chemical Physics 11,* Springer-Verlag, Berlin, Heidelberg, and New York, pp. 3–18.

Pindak, R. (1982). Private communication of results.
Leadbetter, A.J., Mazid, M.A., Kelly, B.A., Goodby, J.W., and Gray, G.W. (1979). *Phys. Rev. Lett.* **43**, 630.
Leadbetter, A.J., Mazid, M.A., and Richardson, R.M. (1980). 'Structures of the smectic B, F and H (G) phases of the N-(4-n-alkoxybenzylidene)-4'-alkylanilines and the transitions between them.' In S. Chandrasekhar (ed.), *Liquid Crystals,* Heyden, London, Philadelphia, and Rheine, pp. 65–79.
Levelut, A.-M., Doucet, J., and Lambert, M. (1974). *J. Phys. (Paris)* **35**, 773.
Levelut, A.-M. (1976). *J. Phys. (Paris)* **37**, 51.
Richter, L., Demus, D., and Sackmann, H. (1976). *J. Phys. (Paris)* **37**, 41.
Sackmann, H. (1979). *J. Phys. (Paris)* **40**, 5.

8 The smectic H phase

Introduction

The smectic H phase had a long history of involvement in the confusion at one time surrounding the nomenclature system for smectic polymorphic modifications. The code letter H was originally used by de Vries and Fishel (1972) to describe the smectic phase of N-(4-n-butyloxybenzyl-idene)-4'-ethylaniline (4O.2), a phase which was later shown (Richter, Demus, and Sackmann, 1976) to be of the same class as one of the smectic phases of a pyrimidine derivative which had previously been designated as G by Demus and co-workers (1971). Other workers had tended to use the letter H to denote the three-dimensional tilted smectic B phase, and not the G coding used by the Halle group. Then in 1979, Sackmann designated a phase below the G phase in terephthalylidene-bis-4-n-butylaniline (TBBA) as H in type. In this case, he was using H for the first time to describe the tilted smectic E phase (E_C). Thus, a dual nomenclature had arisen, with some researchers using H to describe the tilted B phase and G to describe the tilted E phase, and others adopting the reverse notation.

The problem was largely resolved in 1980 when Demus, Goodby, Gray and Sackmann proposed the use of a unified nomenclature system. Based on historical events, the system chosen was the one using G to describe the tilted B phase and H to describe the tilted E phase. Therefore in this section, the letter H is used in the context of the tilted smectic E phase.

Although the S_H phase has obviously been known for a long time (particularly in the case of TBBA), it has not been fully investigated by miscibility studies, differential thermal calorimetry, or X-ray studies. Therefore, its structure and properties are still open to detailed investigation. In the following sections, the nature of the H phase is discussed as far as current knowledge of its structure and properties permit.

Structure of the smectic H phase

Most of the X-ray diffraction studies performed on the S_H phase indicate that it has a structure equivalent to that of the S_E phase, except that the

THE SMECTIC H PHASE

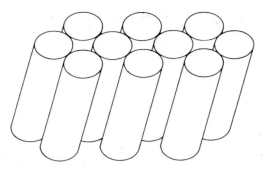

Figure 8.1 The layer structure of a smectic H type phase—the direction of the tilt in relation to smectic H and smectic H' is discussed fully in the text.

molecules have their long axes tilted with respect to the normal to the layer planes. From the description of the phase as being of the S_E type, it must be assumed that the molecules adopt an orthorhombic close-packing in a plane at right angles to the molecular long axes. Hence, because of the tilt, the pseudo-hexagonal net becomes even more distorted, and the phase really has a monoclinic structure, as shown in Fig. 8.1.

It must be assumed that like the S_E, S_G, and correlated S_B phases, the layers are stacked on top of one another in an ordered way, so giving a three-dimensional structure. The smectic phase is therefore of the crystal type. Also, in a similar way to S_F and S_G phases, there will be more than one possible tilt direction of the long axes of the molecules with respect to the orthorhombic net. If the same situation applies as that in G phases, the tilt directions will be towards either a side or an apex of the pseudo-hexagonal net. This would give rise respectively to the phases: S_H—cf. S_F and S_G ($a>b$) and S_H'—cf. S_I and S_G' ($b>a$). Whether S_H and S_H' phases are or are not miscible is not yet known, so at present, different code letters are not used for these two modifications. If they are immiscible, there would be logic in assigning the code letter K to H' (see p. 154).

Gane, Leadbetter, and Wrighton (1981) quote the following cell parameters for the various smectic phases of TBPA and 8OSI.

	TBPA		8OSI
S_F	$a \sim 9.8$ Å $b \sim 5.2$ Å $c = 31.5$ Å $\beta = 113°$	S_I	$a \sim 5.6$ Å $b \sim 9.1$ Å $c = 30.6$ Å $\beta = 113°$
S_G	$a = 9.82$ Å $b = 5.25$ Å $c = 31.6$ Å $\beta = 115°$	S_G'	$a = 5.7$ Å $b = 9.0$ Å $c = 31.0$ Å $\beta = 114°$
S_H	$a = 10.0$ Å $b = 5.3$ Å $c = 30.9$ Å $\beta = 118°$	S_H'	$a = 5.4$ Å $b = 9.1$ Å $c = 31.0$ Å $\beta = 114°$

Volino, Dianoux, and Hervet (1976) have shown from neutron scattering studies that the rotational disorder around the molecular long axes appears to be frozen out in the S_H phase of terephthalylidene-bis-4-n-butylaniline (TBBA). The cell is therefore no longer C-centred, and the only possible disorder is a rotation of π around the long axes. This oscillatory motion is of a similar nature to that observed for the orthogonal S_E phase. The overall structure of the S_H phase is shown in Fig. 8.2 which illustrates the already familiar chevron packing of the molecules.

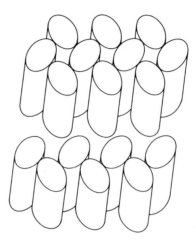

Figure 8.2 Structure of the smectic H phase.

In a similar way to S_C, S_F, and S_G phases, the S_H phase has the elements of tilt required to produce a chiral phase (S_{H*}) provided that the constituent molecules are chiral. As discussed in the previous chapter, results obtained with G^* phases show that these have at least diminished optically active properties in comparison with the precursor S_{C*} and S_{F*} phases. This indicates that the ordered structure makes the formation of the helical arrangement of the tilt directions difficult. Hence, it is even more unlikely that the S_H phase, with its almost crystalline structure, will form a chiral phase. Materials have been prepared that are optically active and show S_{C*}, S_{F*}, and S_{G*} phases, but investigation of the optical properties of the S_{H*} phases formed on further cooling has always proved difficult to perform. Therefore, although it seems unlikely that the S_H phase will exhibit a chiral modification, the possibility cannot be ruled out provided that the inter-layer associations are weak.

Textures of the smectic H phase

The smectic H phase has not yet been obtained directly on cooling the isotropic liquid or the nematic phase, and therefore has not exhibited its natural microscopic texture(s). The phase does however show a number of paramorphotic microscopic textures based on the textures of the phases preceding it on cooling. Commonly, these are the paramorphotic fan and mosaic textures.

The number of compounds for which it has definitely been established that the phase is H as distinct from H' is very small. Consequently, the statistical basis for making comparisons between the textures of S_H and $S_{H'}$ phases is highly insecure. In the following section therefore, no attempt is made to differentiate the textures of smectic H and H' phases since, at present, the two types cannot be distinguished with any certainty on the basis of their microscopic textures. However, the Plates used to illustrate the following discussion of microscopic textures have all been chosen to relate to compounds for which it is known that the phase is either S_H or $S_{H'}$. Similarly, the three standard materials mentioned towards the end of this chapter involve S_H or $S_{H'}$ phases of defined structure.

Paramorphotic textures of the smectic H phase

The focal-conic fan texture can be either broken or of a clear, but strained nature. Usually, the clearer type of fan texture is obtained when the phase is formed in a sequence that involves either an isotropic or a nematic to smectic C transition which produces the clear S_C fan texture. In this case, the smectic C fan texture forms from the preceding phase in the form of bâtonnets. The growth of these bâtonnets (and of the mesophase) involves the deposition of molecules from the preceding phase at the surface of the bâtonnets. Hence, although the phase formed is of the tilted kind, the fan texture is still smooth and clear. In this case, the S_H phase that is subsequently formed in any sequence involving this type of paramorphosis also shows a fairly clear, microscopic fan texture. However, this texture is somewhat strained and shows stress lines running down the backs of the fans. Plate 82 of Sequence 5 shows this clearer focal-conic fan texture plus the mosaic texture of the smectic H' modification of *N,N*-bis-(4'-n-heptyloxybenzylidene)-1,4-phenylenediamine.

It is more common to observe the smectic H phase in its broken fan texture, simply because most phase sequences involve cooling from a smectic A modification. All tilted phases formed on subsequent cooling from the A phase exhibit broken fan textures of one type or another. For the S_H phase, its broken focal-conic form is very similar to that of the S_G phase, i.e. it has a patchwork pattern of chunkily broken areas. At the transition from a S_G to a S_H phase, very little change is therefore actually observed in the fan texture (in comparison with the mosaic textural change), and often the transition can be missed. The broken focal-conic fan texture of $S_{H'}$ is shown in Plate 49, along with the mosaic texture for 4-(2'-methylbutyl)phenyl 4'-n-decyloxybiphenyl-4-carboxylate. There is also very little difference between the focal-conic texture of the achiral S_H phase and the S_H phase formed by optically active materials; Plate 50 shows the paramorphotic broken focal-conic fan texture of the $S_{H'}$ phase of (+)-4-(2'-methylbutyl)phenyl 4'-n-octyloxybiphenyl-4-carboxylate. The texture is identical to that of the racemic modification, indicating that the phase is either only very weakly optically active, or else it has no optically active properties at all.

The smectic H phase can show a number of types of mosaic texture, ranging from small well-defined, clear, platelet types, to large cross-hatched varieties. The transition from the S_G to the S_H phase is probably best characterized by the changes that occur in this particular texture. In the case of N,N-bis-(4'-n-heptyloxybenzylidene)-1,4-phenylenediamine (S_G–$S_{H'}$) the mosaic areas become cross-hatched with a large number of lines, as shown in Plate 81 of Sequence 5. These lines are of a transitory nature and disappear on cooling to give a platelet texture which has smaller platelet areas than those in the preceding $S_{G'}$ phase. The borders of these areas are less smooth than in the preceding mesophase, and show many bends and angles as shown in Plate 82 of Sequence 5 for the fully developed $S_{H'}$ phase. In the case of the 4-(2'-methylbutyl)phenyl 4'-n-alkoxybiphenyl-4-carboxylates, a similar transformation takes place, except that the platelet areas of the $S_{G'}$ phase become crossed with parallel lines (giving a 'corrugated' effect) at the point of transition. On further cooling, these lines again disappear to produce a platelet texture that has smaller mosaic areas than in the preceding phase. The 'lined' transition from the $S_{G'}$ to the $S_{H'}$ phase of 4-(2'-methylbutyl)phenyl 4'-n-octyloxybiphenyl-4-carboxylate (8OSI) is shown in Plate 66 of Sequence 1, and the texture of the resulting $S_{H'}$ phase is shown in Plate 67 of Sequence 1.

The smectic H phase also adopts permanently lined or cross-hatched mosaic textures. Typically, terephthalylidene-bis-4-n-butylaniline (TBBA) shows a classical cross-hatched mosaic texture of the S_H phase. Plate 51 shows the 'zig-zag' cross-hatching lines with which the mosaic S_G texture of TBBA becomes crossed at the transition. These are retained throughout the temperature range of the S_H phase, and are obviously not of a transitory kind. A similar effect can be observed for terephthalylidene-bis-4-n-pentylaniline (TBPA), Plate 52, except that the lines this time are not as pronounced and appear as a graining effect across the mosaic areas. It is interesting to note that in the texture of the S_H phase of this material, the *schlieren* brushes of the preceding S_F phase have been frozen into the texture.

Identification and classification of smectic H phases

Microscopic textures

(a) Smectic H phases have not been obtained directly on cooling the isotropic liquid or the nematic phase, and therefore the natural microscopic textures have not yet been observed.
(b) Smectic H phases do, however, exhibit a number of paramorphotic microscopic textures which fall into two categories—the fan and the mosaic textures. Identification of the phase is not particularly easy by observation of the fan texture alone, but it is more readily identified from its mosaic texture and the changes that occur when this texture is formed from the preceding phase. The mosaic texture of a S_H phase often has small platelet areas, the borders of which are usually splintered or 'crinkled'. Sometimes the platelet regions are cross-hatched by 'zig-zag' lines or grainings, and this type of mosaic texture is usually indicative of a smectic H phase. The way in which the H phase develops in the mosaic texture is however a better indication of its presence. Thus, if the mosaic areas become lined at the transition, and this is followed by a change in the mosaic pattern, this is usually good evidence that a smectic H phase is being formed.

Miscibility studies

(a) Miscibility studies are not particularly easy to perform on smectic H materials. It is usual to obtain a depression in the transition temperature from the S_H phase to the higher-temperature phase in binary mixtures, particularly in the region containing approximately 50% by weight of each component. Proof of the existence of a continuum of S_H phase across the diagram of state may therefore be impossible to obtain if the depressed temperatures lead to crystallization before the S_H phase forms. Also, the microscopic textures of the S_H phases of binary mixtures are different from those observed for the pure compounds. With mixtures, there is a general lack of bright colours in the mosaic areas, which tend to be dull and monotone. Also, the edges of the mosaic areas and the hatching lines have a more wizened appearance.
(b) Standard materials that exhibit S_H or $S_{H'}$ phases and may be useful in miscibility studies:

(i) $\text{n-C}_4\text{H}_9-\text{C}_6\text{H}_4-\text{N=CH}-\text{C}_6\text{H}_4-\text{CH=N}-\text{C}_6\text{H}_4-\text{C}_4\text{H}_9\text{-n}$

Terephthalylidene-bis-4-n-butylaniline (TBBA) (standard S_H material that exhibits a cross-hatched mosaic texture).

$$I \rightarrow N \rightarrow S_A \rightarrow S_C \rightarrow S_G \rightarrow S_H$$

(ii) n-C$_8$H$_{17}$O—⟨⟩—⟨⟩—CO.O—⟨⟩—CH$_2$CH(CH$_3$)C$_2$H$_5$

4-(2'-Methylbutyl)phenyl 4'-n-octyloxybiphenyl-4-carboxylate (8OSI) (standard S$_{H'}$ phase formed from a S$_{G'}$ phase—mosaic areas show transitory lines at the point of transition).

$$I \to N \to S_A \to S_C \to S_I \to S_{G'} \to S_{H'}$$

(iii) n-C$_7$H$_{15}$O—⟨⟩—CH=N—⟨⟩—N=CH—⟨⟩—OC$_7$H$_{15}$-n

N,N'-bis-(4-n-Heptyloxybenzylidene)-1,4-phenylenediamine (cross-hatched transition lines at the S$_{G'}$ to S$_{H'}$ transition—also exhibits a clear fan texture).

$$I \to N \to S_C \to S_I \to S_{G'} \to S_{H'}$$

X-ray diffraction pattern

There are a number of problems in trying to obtain well-aligned samples of S$_H$ phases in order to obtain good diffraction patterns. These problems are often related to domain formation. In general however the X-ray diffraction patterns obtained from S$_H$ phases contain many rings/spots as would be expected on the basis of the highly structured nature of the phase.

Differential thermal calorimetry

The results obtained for smectic G to smectic H transitions by differential thermal analysis and differential scanning calorimetry indicate that the transition is usually first order. The enthalpy of transition, however, appears to vary considerably with the type of compound under examination, and can be in the range 0.25 to 1.25 kcal mol^{-1} ($\sim$1 to 5 kJ mol^{-1}) or even higher in some cases.

References

Demus, D., Diele, S., Klapperstück, M., Link, V., and Zaschke, H. (1971). *Mol. Cryst. Liq. Cryst.* **15**, 161.
Demus, D., Goodby, J.W., Gray, G.W., and Sackmann, H. (1980). *Mol. Cryst. Liq. Cryst. Lett.* **56**, 311 and 'Recommendation for the use of the code letters G and H for smectic phases,' in W. Helfrich and G. Heppke (eds.), *Liquid Crystals of One- and Two-Dimensional Order, Springer Series in Chemical Physics 11*, Springer-Verlag, Berlin, Heidelberg, and New York, pp. 31–33.
De Vries, A., and Fishel, D.L. (1972). *Mol. Cryst. Liq. Cryst.* **16**, 311.
Gane, P.A.C., Leadbetter, A.J., and Wrighton, P.G. (1981) *Mol. Cryst. Liq. Cryst.* **66**, 247.
Richter, L., Demus, D., and Sackmann, H. (1976). *J. Phys. (Paris)* **37**, 41.
Sackmann, H. (1979). *J. Phys. (Paris)* **40**, 5.
Volino, F., Dianoux, A.J., and Hervet, H. (1976). *J. Phys. (Paris)* **37**, 55.

9 The smectic I phase

Introduction

The smectic I phase was discovered in early 1978 through the investigations of Richter (1979) at Halle. The work was first reported on by Sackmann (1979a, b, 1980) in the proceedings from different conferences, and later, in detail, by Richter and co-workers (1981). Their investigations centred mainly upon the terephthalylidene-bis-4-n-alkylanilines (the TBAA series). In the case of the nonyl homologue of the TBAA series, they found an extra phase (later shown to be S_I) injected between the S_F and S_C phases. The nonyl and the decyl members

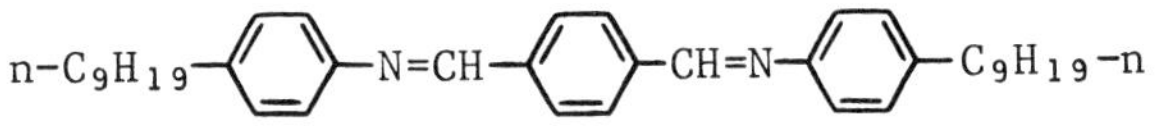

and

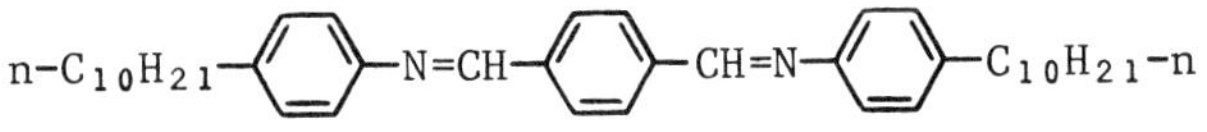

n-C$_9$H$_{19}$—⟨⟩—N=CH—⟨⟩—CH=N—⟨⟩—C$_9$H$_{19}$-n

n-C$_{10}$H$_{21}$—⟨⟩—N=CH—⟨⟩—CH=N—⟨⟩—C$_{10}$H$_{21}$-n

were eventually found to possess the following phase sequence with falling temperature:

$$I \to S_A \to S_C \to S_I \to S_F \to S_G$$

The injection of the S_I phase between S_F and S_C phases was totally unpredicted and unexpected, simply because it was difficult to foresee a smectic structural modification intermediate between that of the S_C phase and that of the S_F phase. Thus, the discovery of the phase in this particular phase sequence led to problems of structural interpretation.

Miscibility studies involving this new phase soon indicated that a number of other materials originally designated as exhibiting tilted S_B phases (by the Halle group) were in fact of the smectic I type, e.g. 4,4'-di-n-octadecyl-oxyazoxybenzene.

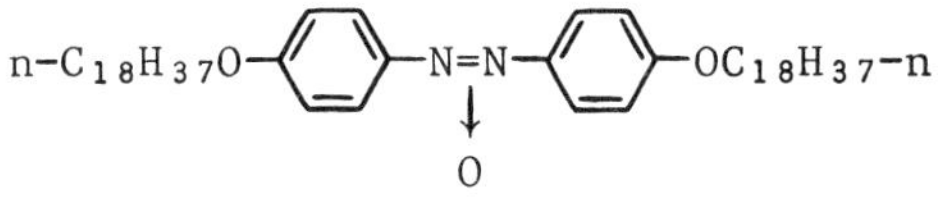

n-C$_{18}$H$_{37}$O—⟨⟩—N=N—⟨⟩—OC$_{18}$H$_{37}$-n
                              ↓
                              O

This compound shows a smectic I phase on cooling the smectic C phase, and not a S_B phase as originally thought.

The same situation arose with the 4,4'-bis-(n-alkylamino)biphenyls

$$\text{AlkylHN}-\underset{}{\bigcirc}\underset{}{\bigcirc}-\text{NHAlkyl}$$

Like the above azoxy compound, the homologue with C_9 alkyl chains forms a S_I phase (and not a smectic B phase) on cooling the S_C phase that is produced from the isotropic liquid.

These reclassifications of S_B phases as S_I phases arose through the discovery that n-pentyl 4-(4'-n-dodecyloxybenzylideneamino)cinnamate

$$C_{12}H_{25}O-\underset{}{\bigcirc}-CH=N-\underset{}{\bigcirc}-CH=CH-CO.OC_5H_{11}$$

is in fact a smectic I material with the phase sequence

$$I \to S_A \to S_C \to S_I$$

The ester also exhibits a monotropic G phase (presumably G'). This ester had been used extensively in earlier miscibility studies as a standard material when its S_I phase was thought to be S_B, and this led to a number of misclassifications. A recent paper by Richter and co-workers (1982) fully describes the corrections of earlier reports necessitated by the various reclassifications.

Also, preliminary indications by personal communication from Richter (1980) that N,N'-bis-(4'-n-heptyloxybenzylidene)-1,4-phenylenediamine (7OBPD) might also exhibit the S_I phase, and not the S_F phase previously reported, have been shown to be correct by Gane, Leadbetter, and Wrighton (1981). In this case, the S_I phase occurs between a S_C phase and a $S_{G'}$ phase.

Similarly, the ester now known as 8OSI exhibits the S_I phase. Originally the phase was thought to be S_F, but X-ray studies by Gane and co-workers (1981) showed that the phase has the same tilt characteristics as the S_I phase of TBDA (terephthalylidene-bis-4-n-decylaniline). This was also the case for both the chiral and the racemic modifications of 4-(2'-methylbutyl)phenyl 4'-n-octyloxybiphenyl-4-carboxylate (8OSI)

$$n-C_8H_{17}O-\underset{}{\bigcirc}\underset{}{\bigcirc}-CO.O-\underset{}{\bigcirc}-CH_2CH(CH_3)CH_2CH_3$$

which have the phase sequence on cooling

$$I \to N \text{ or } Ch \to S_A \to S_C \to S_I \to S_{G'} \to S_{H'}$$

The above X-ray data of Gane and co-workers (1981) relating to the S_I phases of 7OBPD and 8OSI in fact confirmed evidence to this effect obtained by miscibility studies by Goodby (1980). In this context, personal communication from Demus (1982) has shown that his miscibility studies also confirm the X-ray evidence that the phase following the S_C phase in 8OSI is S_I in type.

In the light of their investigations of systems such as 8OSI and TBDA, Gane and co-workers (1981) also pointed out that it seemed clear that the smectic III phase observed by Doucet and co-workers (1978) for the ferroelectric liquid crystal material (HOBACPC)

$$n-C_6H_{13}O-\langle\bigcirc\rangle-CH=N-\langle\bigcirc\rangle-CH=CH-CO.OCH_2CH(CH_3)Cl$$

is also S_I, and this has been confirmed by personal communication from Pindak (1982).

Structure of the smectic I phase

The smectic I phase is a tilted biaxial phase. Powder diffraction photographs show a single outer ring similar to that shown by S_B phases and more sharp than that for the S_F phase. Using an aligned sample of the phase which occurs in the ester

$$n-C_{12}H_{25}O-\langle\bigcirc\rangle-CH=N-\langle\bigcirc\rangle-CH=CH-CO.OC_5H_{11}-n$$

Diele and co-workers (1980) obtained diffraction photographs showing a ring with six maxima when the beam was parallel to the molecular long axes. With the beam parallel to the layers, bars of scattering were observed. Thus the diffractograms were very similar to those of S_F phases.

Gane and co-workers (1981) then took up the problem of the relationship of S_I and S_F phases, and how a structural modification could exist between the S_C and the S_F phase. They first examined TBDA, overcoming problems of obtaining aligned samples by slowly cooling from the isotropic liquid with the sample in a magnetic field of 2T. The diffraction patterns were found then to change reversibly at the S_F to S_I transition at 150°C. They realised that the diffraction patterns arose from tilted pseudo-hexagonal packing arrangements of the molecules with different tilt directions. The situation is best summarized by the following section quoted from the paper:

> For the S_F, the hexagon is tilted towards an edge (or alternatively adjacent 100 rows are uniformly displaced), while for S_I the tilt is towards an apex of the hexagon. In both cases there is no significant distortion from hexagonal packing

in the plane perpendicular to the molecules (the $hk0$ plane of the reciprocal lattice has ~hexagonal symmetry) and the unit cell is C-centred monoclinic with parameters as follows

S_F $a=9.9$ Å $b=5.4$ Å $c=41.6$ Å $\beta=112°$
S_I $a=5.8$ Å $b=9.2$ Å $c=41.7$ Å $\beta=114°$

The scattering centred on the zero level reciprocal lattice nodes in both cases consists of diffuse bars. [The intensity profiles of these] are qualitatively similar for the two phases, showing that both consist of uncorrelated layers (but with long range 3D bond orientational order) and have limited positional correlations in the layers.

Thus, for TBDA, it was shown that the essential difference between the S_F and the S_I phases is one of tilt direction of the pseudo-hexagonal molecular packing. Therefore, with the alteration of tilt direction shown in Fig. 6.2 of Chapter 6, the structure of the S_I phase is in fact that shown in Fig. 6.1 for the S_F phase.

As also pointed out in Chapter 6 on the S_F phase, differences in the in-plane correlation length also arise—those within a layer in the S_I phase being considerably greater. Indeed, Benattar and co-workers (1980) indicated that the S_I phase may be a 2D *crystal,* i.e. the in-plane correlation length was thought to be *very* long. This conclusion has been disputed by detailed X-ray studies of free-standing films of the S_I phase by Budai and co-workers (1984) who have shown that the S_I phase is truly hexatic in nature—see also Chapter 10. However, the main distinguishing feature between S_F and S_I phases is the direction of the tilt.

Gane and co-workers (1981) then went on to study 8OSI and showed that the phase originally thought to be S_F has the same X-ray characteristics and the same tilt direction as that of the S_I phase of TBDA.*

These authors have also considered the profiles of the bars (in fact cylinders) of scattering associated with the $hk0$ reflections in more detail for both the S_F and S_I phases of TBDA. This has been done by calculating the curves for the elastic scattering for completely uncorrelated layers on the basis of the molecule in its most extended conformation and with the appropriate tilt angle. Even allowing for imperfect alignment (mosaic effect), the experimental profile is broader than the calculated profile for both phases. An improved agreement was obtained using the scaled diffuse scattering found for the S_B phase of the nO.m compounds, and this suggested that the bars of scattering have a similar origin and contain a major component connected with molecular motions—to be expected in a system involving weak interlayer associations and disordered terminal alkyl chains.

---

*Note however that as early as 1978, Leadbetter and his colleagues had realised that two tilt directions were possible in these tilted 2D phases (S_I and S_F), but at that time both phases were still denoted as F.

Textures of the smectic I phase

The smectic I phase has been shown to exhibit a natural texture in materials of the type 4,4'-bis-(n-octadecylamino)biphenyl, where the S_I phase is formed directly from the isotropic liquid (see Plate 53). However, very few materials exhibit this direct transition from the isotropic liquid, and it is more often the case that the phase is observed as part of a sequence in which it is formed on cooling an S_A or an S_C phase (see Plate 54). In this situation, the major problem is distinguishing the S_I phase from its close relative the S_F phase.

The S_I phase *can* be distinguished from the S_F phase by optical microscopy. The S_I phase exhibits two main textures, the broken focal-conic fan and the *schlieren* textures. The broken focal-conic fan texture is almost identical with that of the S_F phase (see Chapter 6). However, the *schlieren* texture affords some small differences which can enable one to distinguish between the two phases. In the case of the S_I phase, the *schlieren* texture exhibited shows typical *schlieren* brushes arising from point singularities (see respective smectic C sections); for the S_F phase, the *schlieren* areas are still retained, but the texture is now crossed with fine mosaic lines giving a *schlieren*-mosaic texture. This is emphasised in Sequence 6 for terephthalylidene-bis-4-n-decylaniline (TBDA). Plate 83 shows the S_A phase in its fan and homeotropic textures. On cooling, this phase forms a smectic C phase, Plate 84; further cooling produces the S_I phase which shows a broken fan and *schlieren* texture, Plate 85. Finally, this phase forms the S_F phase on further cooling (Plate 86). It is interesting to compare the two Plates 85 and 86; the difference between the two textures appear quite pronounced, but, taken out of the context of a cooling sequence, it would be difficult to distinguish one phase from the other.

In some cases, the change that occurs at this transition is considerably more pronounced. For terephthalylidene-bis-4-n-nonylaniline (TBNA), the *schlieren* texture of the S_I phase becomes crossed with a mosaic network which fluctuates before settling to give the *schlieren*-mosaic texture of the S_F phase. This change in texture is readily observed, and precisely reversible for TBNA. However, for TBDA the transition is less well defined and appears to occur *via* a process of setting of the *schlieren* into a *schlieren*-mosaic texture.

The distinction between the S_I and S_F phases is obviously very difficult. However, optical microscopy does afford some possible ways of distinguishing between the two phases, *provided* that they are both present in a particular phase sequence. The distinction between S_I and S_F phases occurring separately in other phase sequences is much more problematical, but again distinction between the two phases is best made by observation of the *schlieren* or *schlieren*-mosaic textures.

Identification and classification of the smectic I phase

Microscopic textures

(a) The easiest way to distinguish between the S_I and the S_F phases is to examine carefully the textures formed on cooling the homeotropic texture of the preceding S_A phase or the *schlieren* texture of the preceding S_C phase (S_I is formed from S_C or S_A, and S_F from S_A, S_C or S_I on cooling). If the resulting texture is of a *schlieren* form, which is difficult to bring into microscopic focus, it is more likely to be that of the S_I phase. If the texture is of a mosaic type it is probable that the phase is S_F in nature. Both S_I and S_F phases are readily distinguished in these textural forms.

(b) The S_{I^*} phase (optically active version of S_I) is also identifiable from its bubble texture, see Plate 55 for (+)-4-(2'-methylbutyl)phenyl 4'-n-octyloxybiphenyl-4-carboxylate. The bubbles (or pancakes) change size with changing temperature and flow over one another in a similar fashion to eddies on the surface of a pool of water.

Miscibility studies

(a) In miscibility studies between S_F and S_I materials in which one material exhibits the S_I phase, whilst the other has the S_F phase, apparent co-miscibility can be observed. However, there is always a small percentage region, of the order of 2% of the mixture composition, wherein a definite, reversible S_I to S_F change occurs. This region can be easily missed because of its narrowness, and this occurred in the earlier studies of 8OSI which was initially designated 8OSF.

(b) Standard materials that exhibit the smectic I phase and are useful in miscibility studies:

(i) $C_{10}H_{21}$—⟨⟩—N=CH—⟨⟩—CH=N—⟨⟩—$C_{10}H_{21}$

Terephthalylidene-bis-4-n-decylaniline (TBDA) (a S_I phase occurring between S_F and S_C).

$$I \to S_A \to S_C \to S_I \to S_F \to S_G$$

(ii) $C_8H_{17}O$—⟨⟩⟨⟩—CO.O—⟨⟩—$CH_2CH(CH_3)C_2H_5$

4-(2'-Methylbutyl)phenyl 4'-n-octyloxybiphenyl-4-carboxylate (8OSI) (the S_I phase is optically active if the -$CH_2CH(CH_3)C_2H_5$ group is chiral).

$$I \to N \to S_A \to S_C \to S_I \to S_{G'} \to S_{H'}$$

(iii) $C_{18}H_{37}HN$—⟨phenyl⟩—⟨phenyl⟩—$NHC_{18}H_{37}$

4,4'-bis(n-Octadecylamino)biphenyl (the S_I phase is formed directly from the isotropic liquid).

$$I \rightarrow S_I$$

X-ray diffraction pattern

The X-ray diffraction scattering patterns are discussed in the section on structure of the S_I phase.

Differential scanning calorimetry

DSC and DTA indicate that the S_C to S_I transition is second order or weakly first order (0.3–0.6 kcal mol^{-1}, 1.2 to 2.4 kJ mol^{-1}). The S_I to S_F transition is second order and an order of magnitude smaller in enthalpy value.

References

Benattar, J.J., Moussa, F., and Lambert, M. (1980). *J. Phys. (Paris)* **41**, 1371.
Budai, J., Davey, S.C., Pindak, R., and Goodby, J.W. (1984). *J. Phys. (Paris) Lett.*, to be published.
Demus, D. (1982). Personal communication.
Diele, S., Demies, D., and Sackmann, H. (1980). *Mol. Cryst. Liq. Cryst. Lett.* **56**, 217.
Doucet, J., Keller, P., Levelut, A.-M., and Porquet, P. (1978). *J. Phys. (Paris)* **39**, 548.
Gane, P.A.C., Leadbetter, A.J., and Wrighton, P.G. (1981). *Mol. Cryst. Liq. Cryst.* **66**, 247.
Goodby, J.W. (1980). Unpublished results.
Sackmann, H. (1979a). *J. Phys. (Paris)* **40**, 5.
Sackmann, H. (1979b). Proceedings of the Third Liquid Crystal Conference, Budapest, later published (1981) in L. Bata (ed.) *Advances in Liquid Crystal Research and Applications*, Pergamon Press, Oxford and New York, and Akademiai, Budapest, Vol. 1, pp. 27–38.
Sackmann, H. (1980). 'The system of nonamphiphilic smectic liquid crystals with layer structures.' In W. Helfrich and G. Heppke (eds.), *Liquid Crystals of One- and Two-Dimensional Order, Springer Series in Chemical Physics 11*, Springer-Verlag, Berlin, Heidelberg, and New York, pp. 19–30.
Richter, L. (1979). Dissertation, University of Halle.
Richter, L. (1980). Personal communication.
Richter, L., Demus, D., and Sackmann, H. (1981). *Mol. Cryst. Liq. Cryst.* **71**, 269.
Richter, L., Sharma, N.K., Skubatz, R., Demus, D., and Sackmann, H. (1982). *Mol. Cryst. Liq. Cryst.* **80**, 195.

10 Some new developments in phase classification and structure

Introduction

The science of liquid crystals is rapidly developing and changing, and this makes it rather liable to new theories and experimental studies. Recent investigations are a testimony to this fact, for new experimental techniques—particularly in structural analysis—have brought about many exciting developments and destroyed some widely-held beliefs about certain phases. For example, the use of high-power, high-resolution radiation coupled with new alignment methods has provided a very powerful probe for structural studies.

Since this text was initiated, there have been a number of forward leaps in our knowledge of smectic phases, and some of these more recent developments are reported in this final chapter.

Hexatic and crystal smectic phases

Liquid crystal phases, and smectic phases in particular, have provided a scenario for physical studies of phase transitions, melting behaviour, and low dimensionality. It is not surprising therefore that smectic liquid crystals have recently been used as a means of probing the mechanism of melting in two dimensions. These studies have been elegantly carried out by investigating the properties of free-standing films of liquid crystals which provide monodomain, single crystal specimens of quality. High resolution X-ray diffraction and synchrotron radiation experiments have then been used to facilitate the detailed study of the melting and transition phenomena associated with a number of smectic phases. These results have led to the discovery of a new smectic phase, 'hexatic B', and to a better understanding of the structures of some of the other smectic modifications.

Earlier, Mermin (1968) had shown that for an ideal two-dimensional system, the positional order parameter is zero at all finite temperatures, but that a two-dimensional solid may have directional, orientational long-range order.

Halperin and Nelson (1978) further developed ideas of melting in two dimensions by producing a model based on the concept that the two-dimensional melting process was a dislocation-mediated, second-order transition. For this mechanism, they found that a two-dimensional phase having algebraically decaying bond-orientational order (a hexatic phase) would occur between the two-dimensional solid and the liquid phase. Bond-orientational order refers in this case to the orientation of a bond between two nearest neighbouring atoms relative to a given axis. If the melting process is characterized by an unbinding of dislocation pairs at a temperature, T_m, then the density of free dislocations above T_m will lead to exponential decay of the translational order parameter. However, orientational ordering persists, in the sense that bond angle correlations decay algebraically. The phase can be described as a liquid crystal, similar to a two-dimensional nematic, but with a six-fold rather than a two-fold anisotropy. At temperatures above T_m, the Frank constant, K_A, although finite, decreases; a point is then reached at which the dissociation of disclination pairs drives a transition into an isotropic phase for which both translational and orientational order are decaying exponentially, i.e. the isotropic liquid.

Subsequently, Birgeneau and Litster (1978) extended the ideas of Halperin and Nelson to three dimensions by suggesting a three-dimensional liquid crystal phase consisting of two-dimensional hexatic layers which interact to produce long-range, three-dimensional bond-orientational ordering, i.e. a stacked hexatic phase. In this phase, the positional ordering of the molecules is short-range within the layers and there are no positional correlations between the layers. The bond-orientational ordering can be pictured in this case as an extensive orientational ordering of the hexagonal packing net of the molecules. Although the positional correlations are only short range, the orientation of the hexagonal matrix is the same over a long range, both within and between the layers—see Fig. 10.1.

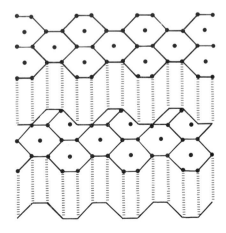

Figure 10.1
Hexatic B structure. The hexagonal net depicts bond-orientational ordering—long range in 3D. Periodicity of the molecules is depicted as ▥▥, indicating their average position.

Classification of hexatic and crystal B phases

At approximately the same time as these hypotheses were being advanced by theoreticians, experimentalists were also noticing differences in the properties of smectic B materials. Byron and co-workers (1980) indicated that the focal-conic textures of the B phases of the N-(4-phenylbenzylidene)-4'-n-alkoxyanilines

$$\text{Ph-Ph-CH=N-Ph-OR}$$

and the 4-(4'-n-alkoxybenzylideneamino)biphenyls

$$\text{Ph-Ph-N=CH-Ph-OR}$$

differed considerably from those of other smectic B materials, notably those of the ester type. These textures were described as stunted focal-conic fan patterns, and were related to a more ordered smectic phase type than that usually envisaged by the term S_B.

Similarly, X-ray diffraction studies by Leadbetter and co-workers (1979a) showed that some B phases had long-range positional ordering of the constituent molecules in three dimensions. These investigations were carried out entirely on the N-(4-n-alkoxybenzylidene)-4'-n-alkylanilines (nO.m's)

$$C_nH_{2n+1}O-\text{Ph}-CH=N-\text{Ph}-C_mH_{2m+1}$$

and the structure of the B modification for these materials was described as a crystal B phase.

Elegant diffraction experiments carried out by Moncton and Pindak (1979) on free-standing films of the smectic B phase of N-(4-n-butyloxybenzylidene)-4'-n-octylaniline (4O.8), confirmed the results of Leadbetter and his co-workers. Consequently, there was a great temptation at this point to claim that all B phases were of a crystalline nature.

However, later in the same year, Leadbetter and co-workers (1979b) demonstrated by X-ray diffraction that two types of B phase clearly existed; one was termed a two-dimensionally ordered smectic phase and the other a three-dimensionally ordered crystal. This observation was subsequently followed by a detailed study of the less ordered smectic B (hexatic B) phase by Moncton and Pindak (1980) and Pindak and co-workers (1981). Using free-standing film techniques, they showed that the material, n-hexyl 4'-n-pentyloxybiphenyl-4-carboxylate (65OBC)

$$C_5H_{11}O-\text{Ph-Ph}-CO.OC_6H_{13}$$

possessed a three-dimensionally stacked, hexatic B phase. The hexatic B phase of this material has short range in-plane correlations of the molecular positions, no interlayer positional correlations, and long-range, three-dimensional, bond-orientational ordering. The S_A–S_B (hexatic) transition was shown to be weakly first-order, or possibly second-order, in nature by detailed heat-capacity studies by Huang and co-workers (1981).

Textural and miscibility studies by Goodby and Pindak (1981) indicated that the crystal B and hexatic B phases should be classified under separate miscibility groups and with different code-letters. However, because of the crystalline nature of the one phase and its standing within a 'liquid' crystal classification scheme, no firm recommendations were made, except to denote one phase as hexatic B and the other as crystal B, and to suggest that in a revised miscibility scheme for smectics, the hexatic B should retain its group letter B, whilst the crystal phase should possess a new code letter L. These suggestions were made in the light of earlier investigations which deemed the B phase to be of a fluid nature with properties similar to the hexatic B phase. Final and conclusive evidence that the two phases have separate identities was obtained by Goodby (1981). The material examined, 4-propionyl-4'-heptanoyloxyazobenzene

$$C_2H_5CO-\phenyl-N=N-\phenyl-O.COC_6H_{13}$$

I ⟵⟶ N ⟵⟶ S_A ⟵⟶ S_B (hex) ⟵⟶ S_B (cryst)   mp
142.5°   141.7°   90.3°   86°   92°

was shown by miscibility methods to possess both types of B phase, with a well-defined transition separating the two. Subsequently, Poeti and co-workers (1982) confirmed Goodby's observations and showed that other members of the same series also exhibited B–B transitions. The transitions were shown to be second order in nature; the enthalpies of 0.02 to 0.05 cal/g, i.e. very small heats of transition, indicated that the phase change is indeed weakly first-order or second-order in nature.

Structural investigations of this phase transition have been made and recently reported in the literature. The first investigation was performed on n-nonyl 4'-n-pentylbiphenyl-4-thiolcarboxylate (95SBC)

$$C_5H_{11}-\phenyl\phenyl-CO.SC_9H_{19}$$

which, it was suggested, might exhibit a B–B transition (Goodby, 1984). The structural features of both phases and the intervening phase transition were reported in detail by Davey and co-workers (1983). Similarly, Albertini and co-workers (1983), continuing studies on the azo-compounds discussed above, reported structural information on the B–B transition.

The role of molecular structure in relation to B phases

The type of B phase exhibited by a particular class of material appears (Goodby, 1984) to be dependent upon the molecular structure, in particular upon the nature of the central linkages employed in the core structure of the molecule. Typically, for a S_B system in which the central aromatic core contains only one linkage, X,

$$\left(\!\!\left\langle\bigcirc\right\rangle\!\!\right)_n \!\!-\!X\!-\!\left(\!\!\left\langle\bigcirc\right\rangle\!\!\right)_m$$

then depending on the chemical nature of this linkage, one of the B phases will be preferred over the other. For example, if the central linkage X is of the Schiff's base type, i.e. $-CH=N-$, then invariably, the B phases of the material are crystalline in nature. The reverse is the case for esters, $X=-CO.O-$, which usually exhibit hexatic B phases. Thiol esters tend to be *predominantly* crystal B in type, whilst azo- and azoxy-materials can exhibit either phase.

These observations were related to the strengths of the lateral dipoles of the central linkages, and it was proposed that the stronger this lateral dipole is, the more likely it is that hexatic phases will predominate. Mesomeric relay of π-electrons through the delocalized portion of the system also appears to have a strong effect. If the carbonyl group of the ester function is not linked into the π-system, a situation which reduces the overall dipole strength, the crystal phases occur in preference to hexatic phases. Thus, where there are strong lateral dipole interactions, there is a tendency for a reduction in the extent of the positional correlation of the molecules. It is suggested that the strong lateral dipoles may create strong attractive interactions which hold the structure of the phase together in a hexatic form, whilst at the same time producing strong repulsions which prevent the phase from collapsing to a crystal. In situations where the lateral dipoles are not particularly strong, then, for the opposite reasons, the crystal B phase dominates.

Textures of the crystal and hexatic B phase

Both phases exhibit mosaic textures when formed directly from the isotropic liquid, and therefore the two phases are indistinguishable from each other in this textural form. However, the two phases also exhibit paramorphotic focal-conic and homeotropic textures when formed by cooling the smectic A phase. The homeotropic texture has been discussed previously in the chapters on smectic A and smectic B phases, and naturally does not provide a means of distinguishing the two phases. The focal-conic fan textures of the two phases and the way these change at the transition from the focal-conic texture of the A phase do however provide distinguishing features necessary for phase identification. Plates 87 to 90 show the textural changes which incorporate the A to hexatic B transition, and these can be compared with Plates 91 to 96, which show the analogous changes for the A to crystal B phase transition. The distinguishing features fall into two categories. Firstly, at the A to B transition, no transition bars are usually observed for the hexatic B case, whereas they are normally observed for the crystal B case. The phenomenon of transition bars was suggested by Goodby and Pindak (1981) to be related to the 3-D structural build-up of the crystal B phase in the A phase at the transition, or its breakdown, depending on the direction of the phase change. This produces a two-phase region which manifests itself as textural bands on the backs of the fans. These transitory bands gradually widen as the transition progresses, then meet and finally coalesce as the change is completed. This development would tend to support the idea of a two-phase region (see Plate 92).

At the A to hexatic B phase change, no transition bars are normally observed. Instead, there is a small change in the birefringence colour and a movement in the radial lines of the focal-conic domains. Reheating the hexatic B to obtain the A phase brings about a greater change; the clear focal-conic texture of the hexatic B phase becomes crossed with parabolic, focal-conic defects—see Fig. 10.2—as the A phase is formed. These defects are presumed to be related to changes in volume occurring at transition; the defects occur at the crystal B to smectic A phase change as well, but affect the texture to a lesser extent. These defects remain in the A phase throughout the temperature range of the mesophase, but near the clearing point they are annealed out, and the fan again becomes smooth—see Plate 90.

Secondly, the two phases are distinguishable by the shapes of the focal-conic domains. These domains are still focal-conic in nature for the hexatic phase, but the conics are altered in the crystal B phase. As the crystal B phase has long range ordering of the molecules in three dimensions, then the layers are not curved or bent easily. Thus, in the fan domains of this phase, the conic sections become angled instead of curved. This can be seen by comparing Plate 88 for the hexatic phase of n-hexyl 4′-n-pentyloxybiphenyl-4-carboxylate with Plates 93 and 96 for the crystal B phase of two analogous n-alkyl 4′-n-pentylbiphenyl-4-thiolcarboxylates. Typically, the crystal B phase is characterized by stepping patterns across the backs and along the edges of the fans. These are possibly related to the layer correlation length for this phase.

Tilted hexatic and crystal phases (I, F, G, and J phases)

Similarly to the orthogonal hexatic B and crystal B phases, there are tilted analogues; these are classified in the miscibility groups I, F, G, and G′ (J).

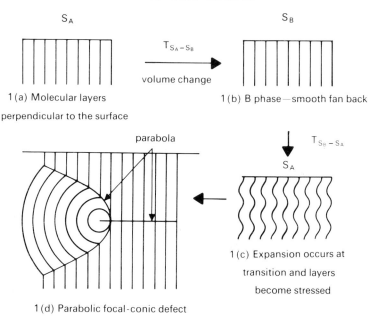

Figure 10.2 Smectic A⟷hexatic B phase change. 1(a)→1(b)—volume contraction has little effect. 1(b)→1(c)/(d)—volume expansion causes localized stresses and textural change. Redrawn from Goodby and Pindak, *Mol. Cryst. Liq. Cryst.* (1981).

Smectic F (Chapter 6) is truly a tilted hexatic phase possessing short-range positional correlations within the layers, no correlation between the layers, and long-range 3-D bond-orientational ordering. As noted earlier (Chapter 9), the absolute structure of the I phase is still in debate. The current belief is that the in-plane correlation length for the S_I phase is long-range, and hence the phase can be described as stacks of two-dimensional crystals. However, there is still the possibility that the I phase is a tilted hexatic phase.

The tilted crystal phases which are the analogues of the orthogonal crystal B phases are the G and G' (J) phases (Chapter 7). Recently, Gane and co-workers (1983) showed that a transition appears to be possible between the G and J phases on the basis of miscibility studies. Hence, the two phases have separate identities and should have different miscibility code letters. The letters suggested for this purpose are G (crystal analogue of F) and J (crystal analogue of I). Crystal G has long-range positional ordering in 3-D, with the molecules tilted with respect to the layer planes. At right angles to the tilt direction, the molecules are hexagonally close-packed, with the tilt direction being towards the side of the hexagonal net. The other suggested miscibility group is J (crystal analogue of I). Crystal J has the same structural features as G, except for the tilt direction, which is now towards the apex of the hexagonal matrix.

Recently, a J to G transition has been observed in the racemic material 4-(2'-methylbutyl)phenyl 4'-n-octylbiphenyl-4-carboxylate

$$C_8H_{17}-\text{\textlangle O\textrangle}-\text{\textlangle O\textrangle}-CO.O-\text{\textlangle O\textrangle}-CH_2\underset{\underset{CH_3}{|}}{C}HC_2H_5$$

which exhibits the following phase sequence

$$I \rightarrow N \rightarrow S_A \rightarrow S_C \rightarrow S_I \rightarrow S_J \rightarrow S_G$$

and confirms that the two phases J and G have separate identities (Budai and co-workers, 1983).

Thus, as with the I and F phases, both of which can occur in a pure material, so it is with the J and G phases, and this makes the definition of a new miscibility group $-J-$ necessary.

Identification and classification of crystal and hexatic B phases

Materials which exhibit hexatic B and crystal B phases and which may be useful in miscibility studies:

(i) $C_5H_{11}O-\text{\textlangle O\textrangle}-\text{\textlangle O\textrangle}-CO.OC_6H_{13}$

$$I \rightarrow S_A \rightarrow S_B \text{ (hex)} \rightarrow S_E$$

n-Hexyl 4'-n-pentyloxybiphenyl-4-carboxylate (65OBC)

(ii) $C_4H_9O-\text{\textlangle O\textrangle}-CH=N-\text{\textlangle O\textrangle}-C_8H_{17}$

$$I \rightarrow S_A \rightarrow SB \text{ (cryst)} \rightarrow Cryst$$

N-(4-n-Butyloxybenzylidene)-4'-n-octylaniline (4O.8)

(iii) $C_2H_5CO-\text{\textlangle O\textrangle}-N=N-\text{\textlangle O\textrangle}-O.COC_6H_{13}$

$$I \rightarrow N \rightarrow S_A \rightarrow S_B \text{ (hex)} \rightarrow S_B \text{ (cryst)}$$

4-Propionyl-4'-n-heptanoxyloxyazobenzene

(iv) $C_8H_{17}O-\text{\textlangle O\textrangle}-\text{\textlangle O\textrangle}-CO.O-\text{\textlangle O\textrangle}-OC_5H_{11}$

$$I \rightarrow N \rightarrow S_A \rightarrow S_C \rightarrow S_B \text{ (hex)}$$

4-n-Pentyloxyphenyl 4'-n-octyloxybiphenyl-4-carboxylate

(v) $C_5H_{11}O$—⟨⟩—⟨⟩—$CO.SC_9H_{19}$

$I \to S_A \to S_B \text{ (hex)} \to S_B \text{ (cryst)} \to S_E$
n-Nonyl 4-n-pentyloxybiphenyl-4-thiolcarboxylate (95SBC)

Materials which exhibit J and G phases and which may be useful in miscibility studies:

(i) $C_6H_{13}O$—⟨⟩—$CH=N$—⟨⟩—$CH=CHCO.OCH_2\overset{*}{C}HClCH_3$

$I \to S_A \to S_{C^*} \to S_{I^*} \to S_{J^*}$
2-Chloropropyl 4-(4'-n-hexyloxybenzylidene)aminocinnamate (HOBACPC)

(ii) $(\pm)C_8H_{17}O$—⟨⟩—⟨⟩—$CO.O$—⟨⟩—$CH_2\underset{CH_3}{CHC_2H_5}$

$I \to N \to S_A \to S_C \to S_I \to S_J \to S_K$
4-(2'-Methylbutyl)phenyl 4'-n-octyloxybiphenyl-4-carboxylate (8OSI)

(iii) C_5H_{11}—⟨⟩—$N=CH$—⟨⟩—$CH=N$—⟨⟩—C_5H_{11}

$I \to N \to S_A \to S_C \to S_F \to S_G \to S_H$
Terephthalylidene-bis-4-n-pentylaniline (TBPA)

(iv) $C_5H_{11}O$—⟨⟩—$CH=N$—⟨⟩—C_7H_{15}

$I \to N \to S_A \to S_C \to S_B \to S_G \text{ (monotropic)}$
N-(4-n-pentyloxybenzylidene)-4'-n-heptylaniline (5O.7)

(v) $C_7H_{15}O$—⟨⟩—$CH=N$—⟨⟩—$N=CH$—⟨⟩—OC_7H_{15}

$I \to N \to S_C \to S_I \to S_J \to S_K$
N,N'-Di-(4-n-heptyloxybenzylidene)-p-phenylenediamine (HEPTOBPD)

Antiphase behaviour

With regard to the smectic A phases of non-cyano-compounds, nothing need be added to what has been said in Chapter 1, but with regard to the smectic A phases of nitro- and cyano-compounds, further information should now be given. Such phases were shown a long time ago, by various workers, to have bilayer structures. Typically, for materials such as the 4-n-alkoxy-4'-cyanobiphenyls, the S_A layer spacing (d) is approximately 1.4 times the actual molecular length (l), estimated with the alkyl chain in its fully extended, all-*trans*-conformation (see Chapter 1). This A phase is sometimes termed S_{A_d}, and the situation for 4-n-octyloxy-4'-cyanobiphenyl is shown below in Fig. 10.3.

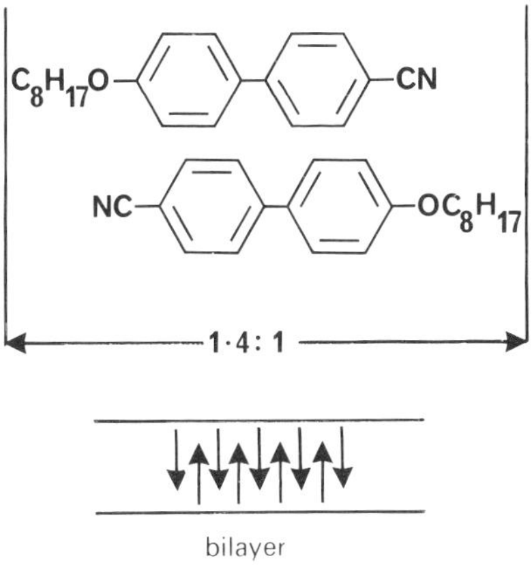

Figure 10.3 Anti-parallel arrangement of molecules of 4-n-octyloxy-4'-cyanobiphenyl in the interdigitated bilayer.

However, this class of material, i.e. one with a strong longitudinal dipole, can also exhibit smectic A phases in which the lamellar spacing is approximately equal to that of the fully extended molecular length. Consequently, it is possible to have A phases that are composed of regions which have a bilayer structure and regions which have a monolayer structure, i.e. a mixture of layering types. These A phases are termed incommensurate (a mixed layered system). The term commensurate is used for a non-mixed system which may be all monolayered or all bilayered.

Generally speaking, these A phases are limited in occurrence to materials which have very strong longitudinal dipole moments associated with powerful electron withdrawing terminal groups, such as nitro- or cyano-substituents. For this phase behaviour, it is advantageous to have this group linked into an aromatic π-electron system which can be easily polarized by the strong electron-withdrawing properties of the terminal substituent. However, if the conjugation between the π-electrons is affected by a lateral dipole group, then a variation on this smectic phase behaviour can be obtained. For example the material

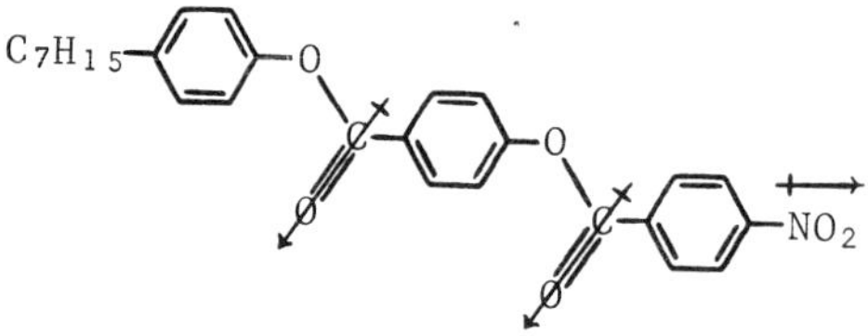

contains this unusual dipole set-up. The nitro-group and carbonyl group of the ester function are both competing for the electrons in the terminal, right-hand aromatic ring. The overall dipolar strength of the two is reduced in comparison with the situation in which the ester function is reversed. However, the material still has a relatively strong terminal dipole (NO_2) and strong lateral dipoles (C=O) giving it dual properties. This effect can be demonstrated by considering the distribution of electrons in the extreme canonical structures (1) and (2) shown in Fig. 10.4 and in the resonance hybrid.

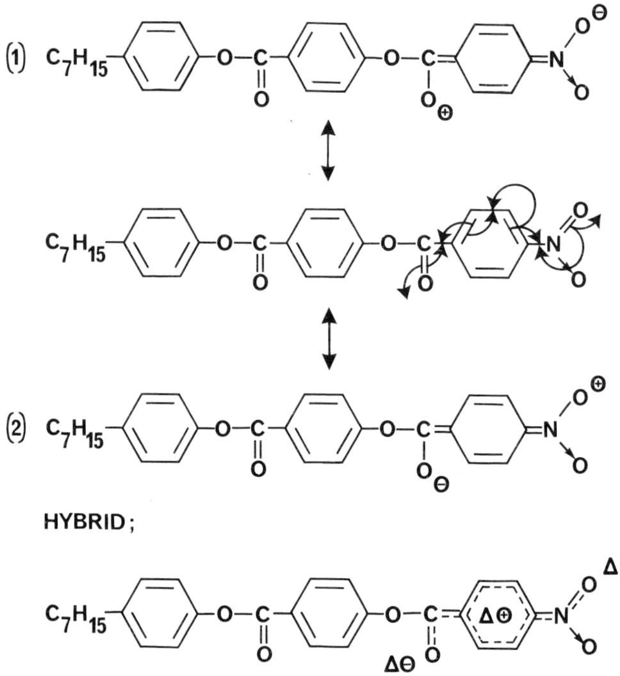

HYBRID;

Figure 10.4 Extreme canonical structures (1) and (2) and the resonance hybrid structure for a diester.

SOME NEW DEVELOPMENTS IN PHASE CLASSIFICATION AND STRUCTURE   145

The net result of interactions between molecules with this type of electronic structure appears to be the formation of a bilayer structure which has a layer spacing that is almost twice the molecular length. The effective overlap between two interacting molecules is then zero or very small, and the difference between a bilayer and a monolayer structure is also rather subtle. In graphic depictions of these phases, the molecules are usually shown as ↑, with the arrowhead representing the terminal electronegative substituent. Two phases can therefore be described by this notation, as shown in Fig. 10.5.

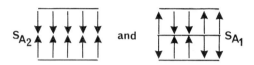

Figure 10.5 Bilayer (S_{A_2}) and monolayer (S_{A_1}) smectic A phases.

S_{A_2} is the bilayer A phase, and S_{A_1} is the monolayer A phase, as shown by Hardouin and co-workers (1980). The S_{A_2} phase is usually the higher-temperature phase, and transitions between S_{A_2} and S_{A_1} have been shown to occur. The heat of transition reported is relatively small however, of the order 1.0 cal/g. The two phases are uniaxial, with a disordered layered structure consisting of molecules or pairs of molecules.

Following from these two phases, a number of other phase structures become possible. For example, the tilted analogues (Fig. 10.6) of S_{A_2} and S_{A_1} are known, and these are called S_{C_2} and S_{C_1}. Both are biaxial and have a disordered packing of the molecules within the layers.

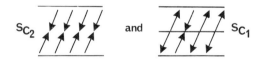

Figure 10.6 Bilayer (S_{C_2}) and monolayer (S_{C_1}) smectic C phases.

Between the S_{A_2} and S_{A_1} phase another phase has been observed; this is called the smectic A antiphase, and was denoted by $S_{\tilde{A}}$ by Hardouin and co-workers (1981). This phase is presumed to have an S_{A_2} type structure, but with a long-wavelength density modulation imposed on the in-plane layer direction. This is shown in Fig. 10.7(a), where the molecules are shown as arrows, although these have been omitted between the maxima and minima of the density modulation. If the eye looks along the bilayer X and follows the modulation, it will see a bilayer structure. However, if the eye looks along the bilayer X in a plane orthogonal to the molecular long axes at the maxima and minima of the modulation, it will see an approximation of the situation shown in Fig. 10.7(b), i.e. bilayer regions of alternating antiparallel polarity—the $S_{\tilde{A}}$ phase.

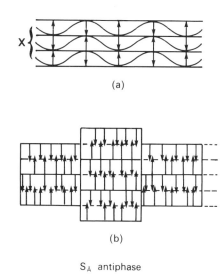

$S_{\tilde{A}}$ antiphase

Figure 10.7 Origin of the antiphase $S_{\tilde{A}}$.

Similarly, it is possible to have a tilted antiphase ($S_{\tilde{C}}$) which is analogous to the S_{C_2} and S_{C_1} family.

This tilted antiphase, called the ribbon phase, has also been shown to exist in materials of the type discussed above. Preliminary structural investigations on this phase indicate that it may be made up of rectangular slabs of bilayer domains. The ribbon or serrated phase has in effect a bilayer structure (Fig. 10.8) that is broken periodically by defect walls. The structures on the two sides of the wall are translated by one-half of the layer thickness, i.e. by about one molecular length. The phase has been shown not to be uniaxial.

This phase is similar in nature to that of the orthogonal antiphase, except that the molecules are tilted, and because of this, it has been called the tilted antiphase, and coded $S_{\tilde{C}}$ by Hardouin and co-workers (1982).

Lastly, there is possibly another phase that can be included in a section dealing with such types of phase phenomena. This is a tilted phase that occurs between the S_{A_1} and S_{C_2} phases on cooling. Its structure has not been elucidated as yet, but it has been observed with a number of different materials by Tinh and co-workers (1982). However, it is possible that this phase is another example of a C antiphase.

SOME NEW DEVELOPMENTS IN PHASE CLASSIFICATION AND STRUCTURE 147

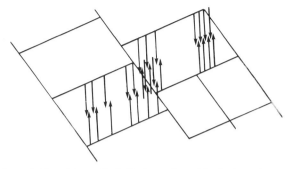

Figure 10.8 Smectic C antiphase ($S_{\tilde{C}_2}$). Redrawn from Hardouin and co-workers, *J. Phys. (Paris) Lett.* (1982).

Materials which exhibit some of these novel phase types:

(i) C_8H_{17}—〈○〉—O.OC—〈○〉—O.OC—〈○〉—NO_2

4-n-Octylphenyl 4-(4'-nitrobenzoyloxy)benzoate (DB_8NO_2)
$I \rightarrow N \rightarrow S_{Ad} \rightarrow S_{\tilde{C}} \rightarrow S_{A_2}$

(ii) $C_{10}H_{21}$—〈○〉—O.OC—〈○〉—O.OC—〈○〉—NO_2

4-n-Decylphenyl 4-(4'-nitrobenzoyloxy)benzoate ($DB_{10}NO_2$)
$I \rightarrow N \rightarrow S_{Ad} \rightarrow S_{\tilde{C}} \rightarrow S_{A_2}$

(iii) $C_8H_{17}O$—〈○〉—CO.O—〈○〉—CH=CH—〈○〉—CN

Trans-4-(4''-Octyloxybenzoyloxy)-4'-cyanostilbene (T8)
$I \rightarrow N \rightarrow S_{Ad} \rightarrow$ re-entrant $N \rightarrow S_{A_2}$

(iv) C_5H_{11}—〈○〉—O.OC—〈○〉—O.OC—〈○〉—CN

4-n-Pentylphenyl 4-(4'cyanobenzoyloxy)benzoate (DB_5CN)
$I \rightarrow N \rightarrow S_{A_2}$

(v) $C_9H_{19}O$—〈○〉—O.OC—〈○〉—O.OC—〈○〉—NO_2

4-n-Nonyloxyphenyl 4-(4'-nitrobenzoyloxy)benzoate (DB_9ONO_2)
$I \rightarrow N \rightarrow S_{Ad} \rightarrow$ re-entrant $N \rightarrow$ re-entrant $S_{Ad} \rightarrow$ re-entrant $N \rightarrow S_{A_1} \rightarrow S_{\tilde{C}} \rightarrow S_{C_2}$

(vi) C$_9$H$_{19}$O—⟨⟩—⟨⟩—O.OC—⟨⟩—CN

4'-n-Nonyloxy-4-biphenylyl 4-cyanobenzoate
I→N→S$_{A_2}$→S$_{\tilde{C}}$→S$_C$

(vii) C$_7$H$_{15}$—⟨⟩—O.OC—⟨⟩—O.OC—⟨⟩—NO$_2$

4-n-Heptylphenyl 4-(4'-nitrobenzoyloxybenzoate) (DB$_7$NO$_2$)
I→N→S$_{A_1}$→S$_{\tilde{A}}$

Textures of antiphases

(a) Typically all orthogonal A phases of cyano-compounds, be they A_2, A_1, or Ã in type, exhibit focal-conic fan and homeotropic textures. Therefore the phases are difficult to distinguish from one another without observing textural changes occurring at the phase transitions. Plates 102–104 show the changes which occur, on cooling, at the smectic A_1 to smectic A antiphase transition for the material 4-n-heptylphenyl 4-(4'-nitrobenzoyloxy)benzoate. The higher temperature A_1 phase is characterised by an angular focal-conic domain structure. Transition occurs with motion of the parabolic focal-conic defects and the radial lines of the domains. The resulting fan texture of the Ã phase is typical of what is regarded as a normal A phase. Thus, the transition appears as an A to hexatic B phase change, but in reverse, i.e. as if a hexatic B type texture were changing to an A type texture on cooling.

(b) The smectic C̃ antiphase phase that occurs between the A_2 and the C phase on cooling compound (vi) above, exhibits a broken focal-conic and spherulitic texture, as shown in Plates 99–101. The clear focal-conic fan texture of the A_2 phase becomes broken on cooling, in a similar fashion to that of the S_F phase. Any homeotropic areas, however, become crossed with a disc or droplet pattern which is reminiscent of the discotic textures of di-isobutylsilanediol. In fact, it is also possible to relate the unusual electronic distributions of these antiphase materials to those of S_D materials (see Chapter 4) which could themselves have discotic characteristics.

Smectic B phases

As with the A phases, A_2, A_1, and Ã, it is possible to extend this phase behaviour in principle to smectic B. For example, the smectic B phase of *trans,trans*-4-n-propylbicyclohexyl-4'-carbonitrile has been shown by Brownsey and Leadbetter (1981) to exhibit a bilayer structure—see Fig. 10.9.

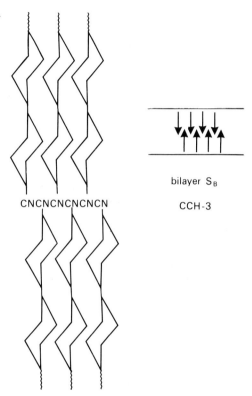

Figure 10.9 Bilayer S_B structure of *trans,trans*-4-n-propylbicyclohexyl-4'-carbonitrile (CCH-3).

This type of structuring is very similar to that of the S_{A_2} phase, except that the molecules are hexagonally close-packed. Thus, this phase would qualify as the equivalent of S_{B_2}. Certainly, the B phase of this material is *not* miscible with conventional B phases of monolayer structuring. Presumably, the injection of discrete molecules of the latter type will split the pairing of the molecules of the cyano-material and so rapidly reduce the chances of observing co-miscibility.

Ferroelectric phases

All of the tilted smectic phases can exhibit ferroelectric properties when their constituent molecules are themselves optically active. The C, I, F, G,

G'(J), H, and H'(K) modifications all fall into this category. However, there is a subdivision of this group of phases based on whether the phase is crystalline in nature. The C, I, and F phases are true smectic liquid crystal phases; the remainder are smectic crystal phases.

In the smectic liquid crystal variety, the phase itself is optically active as a result of the arrangement of the individual chiral molecules. This structural asymmetry takes the form of a twist of the tilt direction on moving from layer to layer about an axis perpendicular to the layers. The individual phases are distinguishable from each other by the in-plane structuring (see the appropriate chapters on C, I, and F phases). An individual layer is therefore ferroelectric, but because of the macromolecular helical arrangement, the bulk phase is not. On the other hand, the crystal phases do not possess this helical arrangement, because the crystal-like structure suppresses the formation of a helix. This results in the formation of polydomains. Each individual domain is therefore ferroelectric, but as with the less ordered smectics, the bulk phase is not, because of its polydomain structure in which every conceivable orientation of the molecules is represented. This leads to the interesting possibility that an *aligned* crystal smectic phase would indeed be ferroelectric. Thus the G, G' (J), H and H' (K) phases are non-chiral ferroelectrics when their constituent molecules themselves are optically active.

Doucet and co-workers (1978) were the first to demonstrate that the crystal G'(J) phase of R(-)-2-chloropropyl 4-(4'-hexyloxybenzylidene)aminocinnamate (HOBACPC) has no helical structure. For this material

$$C_6H_{13}O-\text{C}_6H_4-CH=N-\text{C}_6H_4-CH=CHCO.OCH_2\overset{*}{C}HClCH_3$$

they reported the following phase sequence

$$\text{Cryst} \xrightarrow{60°} S_4 \xleftrightarrow{68°} S_3 \xleftrightarrow{74°} S_{C^*} \xleftrightarrow{81°} S_A \xleftrightarrow{130°} I$$

where S_3 is in fact S_I, and S_4 is $S_{G'}$ (S_J). The phases S_3 and S_4 were shown to be structurally very similar, with their constituent molecules tilted and hexagonally close-packed in layers. The tilt direction of both phases was towards the apex of the hexagonal net. The two phases differ by the extent of the interlayer correlations, S_4 being a correlated, three-dimensional ordered phase, and S_3 retaining the fluid-like layers. Doucet and co-workers however did not classify either of these phases; this was done at a later date, but, because of the changes in the classification scheme for smectics, the letters used are now out-dated. Gane and co-workers (1981) studied the phase behaviour of this material further, and report that its phase sequence is

$$\text{Cryst} \rightarrow S_{J^*} \leftrightarrow S_{I^*} \leftrightarrow S_{C^*} \leftrightarrow S_A \leftrightarrow I$$

SOME NEW DEVELOPMENTS IN PHASE CLASSIFICATION AND STRUCTURE

where S_{C*} and S_{I*} are chiral, optically active liquid crystal phases and S_{J*} is a non-chiral phase, even though it still has the optical activity arising from the chiral nature of the molecules of the material itself. Subsequently, the J and G phases of 4-(2'-methylbutyl)phenyl 4'-n-octyl-biphenyl-4-carboxylate (8SI)

$$C_8H_{17}\text{-}\phi\phi\text{-}CO.O\text{-}\phi\text{-}CH_2\overset{*}{C}HC_2H_5$$
$$|$$
$$CH_3$$

$$I \leftrightarrow Ch \leftrightarrow S_A \leftrightarrow S_{C*} \leftrightarrow S_{I*} \leftrightarrow S_{J*} \leftrightarrow S_{G*}$$

were also shown to be non-chiral by Budai and co-workers (1983). Although we can never be certain that crystal forces will always unwind the helical structure of every ordered, tilted, chiral smectic, it appears that all the crystal smectic phases of this type that have been examined are non-chiral, i.e do not possess a helical arrangement of the tilt direction. However, as already mentioned in Chapter 7, in mixed systems there may be enough disruption of the layer structure of the 'soft' crystal phase to allow adoption of a helical structure.

Clearly, this information is important with respect to the commercial applications of these unusual phases. Clark and Lagerwall (1980) have recently demonstrated a fast-switching, bistable electro-optic effect based on the ferroelectric properties of smectic liquid crystals. Apart from the switching speed, the most important aspect of their concept is that the device is bistable. The bistable operation of this device depends on the suppression of the helix of the chiral phase by surface forces. Therefore, the thickness of the cell has to be less than the pitch-length of the phase. Early results by Cladis and co-workers (1983) show that this is not however always the case, and that bistable operation can only occur in devices of 1–3 µm thickness. However, the crystal phases J and G switch in a truly bistable manner, because they do not possess this helical arrangement.

The smectic F to isotropic liquid transition

The smectic F phase is one of the few phases whose formation had not been observed on cooling directly from the isotropic liquid. Recently however, some of the 4-n-alkanoyloxy-4'-n-octyloxybiphenyls

$$C_8H_{17}O\text{-}\phi\phi\text{-}O.OCC_nH_{2n+1}$$

were shown by Walton and co-workers (1984) to exhibit isotropic-S_F phase transitions. Unfortunately, these transitions are soon followed by formation of a smectic G phase on cooling, so that the F phase exists for only a degree or so. Effectively, the transition from the isotropic liquid becomes $I-S_F/S_G$. However, in the case of binary mixtures with terephthalylidene-*bis*-4-n-pentylaniline (TBPA; N, A, C, F, G, and H phases), isotropic liquid to S_F phase transitions with a wider temperature range F phase are observed before transition to the G phase occurs.

The texture of the F phase as it separates from the isotropic liquid is shown in Plate 105 for 4-n-dodecanoyloxy-4'-n-octyloxybiphenyl. This phase forms in spherical droplets which have a hexagonal cross of optical discontinuity centred in them. The hexagonal discontinuity is possibly related to the hexagonal structure of the phase, whilst the droplet nature is related to the fluidity of the phase.

Structural features of smectic phases

Smectic phases are normally classified by miscibility methods and it is the coding developed from these experiments that has been used to describe all of the smectic modifications known today. However, though based upon miscibility criteria, there has been a tendency to relate this classification scheme to one based on the *structures* of the phases. Many of the problems of phase classification have in fact arisen because of the neglect of one of the methods of study, be it miscibility or structural. Total classification of a phase can only be achieved by the combined use of a number of methods which include microscopy, miscibility, thermal analysis, and X-ray diffraction. With the development of new and powerful X-ray diffraction techniques, it has been possible to assign a certain number of structural features to each individual miscibility group, and the authors consider that it would be useful to give these for the reader's reference in Tables I, II, and III, and in Fig. 10.10. For further reading in the area covered by Tables I to III, the reader is referred to J.W. Goodby (1983) *Mol. Cryst. Liq. Cryst. Lett.* **92,** 171.

SOME NEW DEVELOPMENTS IN PHASE CLASSIFICATION AND STRUCTURE 153

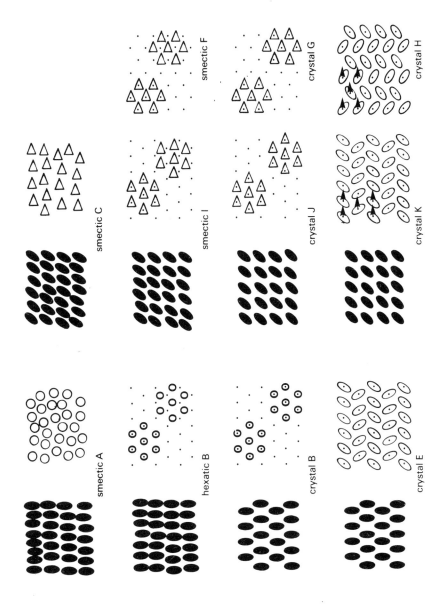

Figure 10.10 Side elevation and plan representations of the molecular ordering in each of the smectic modifications. Triangles or arrows are used to represent tilt direction.

Table I Some structural features of smectic liquid crystal and crystal phases

Miscibility group	Phase type	Molecular packing	Tilt disposition	In-plane correlation	Layer correlation	Bond-orientational order	Dimensionality (ordering)
A	fluid	random	orthogonal	short-range	none	short-range	—
B (hexatic)	hexatic	hexagonal	orthogonal	short-range	none	long-range	3D hexatic
C	fluid	random	tilted	short-range	none	short-range	—
D	plastic	micellar? rods?	—	long-range	—	—	isotropic
E	crystal	orthorhombic	orthogonal	long-range	long-range	long-range	3D
F	hexatic	pseudo-hexagonal	tilt to side	short-range	none	long-range	3D hexatic
G	crystal	pseudo-hexagonal	tilt to side	long-range	long-range	long-range	3D
H	crystal	herringbone, monoclinic	tilted*	long-range	long-range	long-range	3D
I	hexatic 2D crystal	pseudo-hexagonal	tilt to apex	possibly long-range	none	long-range	stacked 2D crystals or 3D hexatic
J(G')	crystal	pseudo-hexagonal	tilt to apex	long-range	long-range	long-range	3D
K(H')	crystal	herringbone, monoclinic	tilted*	long-range	long-range	long-range	3D
L (crystal B)	crystal	hexagonal	orthogonal	long-range	long-range	long-range	3D

*K and H differ as to the side (a or b) of the monoclinic cell towards which the molecules are tilted.

Table II Structural features of the lamellar phases of optically active materials

Phase type and miscibility group		Molecular packing	Layer correlation	In-plane correlation	Tilt direction	Bond-orientational order	Helix structure
smectic liquid crystal phases	C*	random	none	short-range	tilted	short-range	yes
	I*	pseudo-hexagonal	none	possibly† long-range	to apex	long-range	yes
	F*	pseudo-hexagonal	none	short-range	to side	long-range	yes
smectic crystal phases	J*	pseudo-hexagonal	long-range	long-range	to apex	long-range	no
	G*	pseudo-hexagonal	long-range	long-range	to side	long-range	no
	K*	herringbone	long-range	long-range	to side‡	long-range	no
	H*	herringbone	long-range	long-range	to side‡	long-range	no

† May be a 2D crystal or a hexatic.
‡ See footnote to Table I.

Table III Some structural features of bilayer/monolayer A and C phases

Phase type	Layering	Molecular packing	Tilt disposition	Density wave fluctuation
A_2	bilayer	random	orthogonal	—
A_1	monolayer	random	orthogonal	—
$\tilde{A}$	bilayer	random	orthogonal	long-range
Ad	interdigitated bilayer	random	orthogonal	—
C_2	bilayer	random	tilted	—
C_1	monolayer	random	tilted	—
$\tilde{C}$	bilayer	random	tilted	long-range

References

Albertini, G., Melone, S., Poeti, G., Rustichelli, F., and Torquati, G. (1983). *Mol. Cryst. Liq. Cryst.* **99**, 385.
Birgeneau, R.J., and Litster, J.D. (1978). *J. Phys. (Paris) Lett.* **39**, 399.
Brownsey, G.J., and Leadbetter, A.J. (1981). *J. Phys. (Paris) Lett.* **42**, 135.
Budai, J., Davey, S.C., Pindak, R., and Goodby, J.W. (1984). *J. Phys. (Paris) Lett.*, to be published.
Byron, D.J., Keating, D.A., O'Neill, M.T., Wilson, R.C., Goodby, J.W., and Gray, G.W. (1980). *Mol. Cryst. Liq. Cryst.* **58**, 179.
Cladis, P.E., Brand, H.R., and Finn, P.L. (1983) *Phys. Rev.* **A28**, 512.
Clark, N.A., and Lagerwall, S.T. (1980). *Appl. Phys. Lett.* **36**, 899.
Davey, S.C., Fontes, E.Jr., Moncton, D.E., and Pindak, R. (1983). *Stacked Hexatic Phase Transitions.* The proceedings of the 40th Annual Pittsburgh Diffraction Conference, Pittsburgh, USA (November 1982).
Doucet, J., Keller, P., Levelut, A.-M., and Porquet, P. (1978). *J. Phys. (Paris)* **39**, 548.
Gane, P.A.C., Leadbetter, A.J., and Wrighton, P.G. (1981). *Mol. Cryst. Liq. Cryst.* **66**, 247.
Gane, P.A.C., Leadbetter, A.J., Wrighton, P.G., Goodby, J.W., Gray, G.W., and Tajbakhsh, A.R. (1983). *Mol. Cryst. Liq. Cryst.* **100**, 67.
Goodby, J.W. (1981). *Mol. Cryst. Liq. Cryst. Lett.* **72**, 95.
Goodby, J.W., and Pindak, R. (1981). *Mol. Cryst. Liq. Cryst.* **75**, 233.
Goodby, J.W. (1984). 'Relationships between molecular structure and the incidence of crystal B and hexatic B phases.' Presented at the 183rd American Chemical Society Meeting, Las Vegas, USA (March/April 1982). In *Liquid Crystals and Ordered Fluids* (A.C. Griffin and J.F. Johnson, eds.), Vol. 4, Plenum Press, New York, p. 175.
Halperin, B.I., and Nelson, D.R. (1978). *Phys. Rev. Lett.* **41**, 121.
Hardouin, F., Levelut, A.-M., Benattar, J.J., and Sigaud, G. (1980). *Solid State Comm.* **33**, 337.
Hardouin, F., Sigaud, G., Tinh, N.H., and Achard, M.F. (1981). *J. Phys. (Paris) Lett.* **42**, 63.
Hardouin, F., Tinh, N.H., Achard, M.F., and Levelut, A.-M. (1982). *J. Phys. (Paris) Lett.* **43**, 327.

Huang, C.C., Viner, J.M., Pindak, R., and Goodby, J.W. (1981). *Phys. Rev. Lett.* **43,** 630.
Leadbetter, A.J., Mazid, M.A., Kelly, B.A., Goodby, J.W., and Gray, G.W. (1979a). *Phys. Rev. Lett.* **43,** 630.
Leadbetter, A.J., Frost, J.C., and Mazid, M.A. (1979b). *J. Phys. (Paris) Lett.* **40,** 325.
Mermin, N.D. (1968). *Phys. Rev.* **176,** 150.
Moncton, D.E., and Pindak, R. (1979). *Phys. Rev. Lett.* **43,** 701.
Moncton, D.E., and Pindak, R. (1980). 'X-ray studies of smectic B–liquid crystal films.' In S.K. Sinha (ed.), *Ordering in Two-Dimensions,* North Holland Press, New York, p. 83.
Poeti, G., Fanelli, E., and Guillon, D. (1982). *Mol. Cryst. Liq. Cryst. Lett.* **82,** 107.
Pindak, R., Moncton, D.E., Davey, S.C., and Goodby, J.W. (1981). *Phys. Rev. Lett.* **46,** 1135.
Tinh, N.H., Hardouin, F., and Destrade, C. (1982). *J. Phys. (Paris)* **43,** 1127.
Walton, C.R., and Goodby, J.W. (1984). *Mol. Cryst. Liq. Cryst. Lett.* **92,** 263.

Index

A—smectic A   1
A_1—smectic A_1 143, 144
A_2—smectic A_2 145
A_d—smectic A_d 143
$\tilde{A}$—antiphase   145, 146
AB transition   20
algebraic decay   19
alkyl chains   2
alternations of transition temperatures   25
antiphase behaviour   143
arced focal-conics   89, 113

B—smectic B   23
—crystal B and hexatic B   24, 25, 134
B–B transition   137
B_2—smectic B_2 149
B—injection of in miscibility experiments   101
bâtonnets   8
biaxial smectic E   83
biaxiality   46, 129
bilayer smectic A   6, 145
bilayer smectic B   31, 149
bilayer structures   2, 7, 143
birefringence   60
blue phases   xxii
body-centred cubic S_D 70
bond-orientational order   96, 134, 135
broken focal-conics
—S_C 60
—S_F 100
—S_H 123
Brownian motion   64
bubble texture   102

C—smectic C   45
C_1—smectic C_1 146
C_2—smectic C_2 146
$\tilde{C}$—antiphase   146, 147
chevron packing   85, 86, 108
chiral smectics
—C   60
—F   99
—G   110
—H   122
cholesteric phases   xxi
classification of smectics   1, 23
classification—crystal B and hexatic B   136
classification diagram   153
classification tables   154, 155, 156
columnar phases   xxi
cone of revolution   14
conic sections   10, 11, 12
conoscopic figure   36, 46
continuum   19
co-ordinated rotation   28
correlations, between layers   25, 97
correlations, within layers   95, 130, 140
cross-hatched texture   124
crystal phases   121, 134
cubic phase   69, 70, 77, 78
cyano-compounds   7, 72, 143, 147, 148
cyclides (Dupin)   9, 10, 11

D—smectic D   68
dendrites   79
density wave   19
deuterated compounds   6
dielectric anisotropy   7

159

differential scanning calorimetry (DSC)
 —A 21
 —B 43
 —C 66
 —D 80
 —E 93
 —F 104
 —G 118
 —H 126
 —I 133
differential thermal analysis (DTA)—*see* DSC
diffuse layers 5, 6
diffusive motion 109
di-isobutylsilanediol 75
dipole torque 49
dipoles, transverse 49, 53, 56
disc-shaped molecules xxi
discontinuity, optical 8, 9, 15, 101
discotics xxi, 75
domain, focal-conic 15
droplets, smectic A 19
droplet texture of smectic E 88
Dupin cyclides 9, 10, 11
dynamic wave motion 33, 34

E—smectic E 82
electro-optic effect, chiral smectic C 62
ellipse 9, 13
enthalpy 21

F—smectic F 94, 151, 152
fan texture 8
ferroelectric, chiral smectic C 62
ferroelectric, orthogonal 50
ferroelectric phases 149
ferroelectric, tilted 50
first order transition 21
focal-conic 8, 9, 15
focal-conics, deformed 16
focal-conic texture
 —A 15
 —B 38, 139
 —C 60
 —chiral C 64
 —E 89
 —F 100
 —G 113
 —H 123
 —I 131
free rotation 26
free-standing film 134

G—smectic G 23, 105
Grandjean terraces 1, 64
gyroscopic motion 4

H—smectic H 23, 105, 120
helix 61
herringbone structure 85, 86, 108
hexagonal close packing 23
hexagonal net dimension 26
hexatic phases 134
homeotropic textures 8, 35, 42
hyperbola 9, 13

I—smectic I 127
immiscibility 24, 106
impurity effects 19, 39
induced dipoles 53
in-plane correlations 96, 130, 140
inter-layer correlations 25
interstitial hole 31
iridescence 63
iso-S_{AB} transition 40
isotropic texture 78

J—smectic J 110, 140, 151, 153
jointed-rod structure 71, 77

K—smectic K 121, 142, 150, 153
kinking 5

lamellar spacing 2, 4
layer correlation length 98
layer modulation 29, 33, 34, 146
layer stacking 31
layer tilting 34, 35
lined texture, chiral S_C 64
lined texture, mosaic 89, 90
liquid-like chains 2, 3
long-range order 29, 30
lyotropic systems 70

McMillan's model for smectic C 48
melting, in two dimensions 134
micellar phase, S_D 70
miscibility, phase injection 103, 117
miscibility studies 19, 42, 65, 80, 91, 103, 116, 132
modulations, layer 29, 33, 34, 146
molecular columns 29, 30
molecular oscillation 86
molecular rotation 4, 26
molecular structure and B phase type 138
molecular structure, S_C 54
monoclinic cell, S_G 107
mosaic platelets
 —B 36, 37
 —F 101
mosaic texture
 —B 36
 —E 89
 —G 113
 —H 124

INDEX

mosaic-*schlieren* texture
  —F   101
  —I   131
moss texture, smectic B   41, 91
Mössbauer spectra   25

natural texture   35, 88, 112
neutron scattering   2, 107, 108, 122
nitro-acids (S_D), synthesis of   72, 73

optical discontinuity   8, 15, 101
orientational order parameter   4, 5
orthorhombic phase   85
oscillatory motion   86, 87, 122
outboard dipole   49, 53

parabolic focal-conic defects   39, 139
paramorphotic texture   38
patchwork fans   113
petal texture, chiral smectic C   63
phonons   107
pitch, S_C 61
platelets   xxii, 36 37, 89
point singularity   57, 58, 59
polarization   62
polygonal texture   16, 17, 18
polymorphism   1
pretransitional effects   34
pseudo-hexagonal phase   97
pseudo-homeotropic texture   63, 102

re-entrant nematic   8
ribbon phase   146
reorientational motion   86
rotation, frozen   49, 50

schlieren brushes   57, 58
schlieren-mosaic   101, 131
schlieren/schlieren-mosaic   131
schlieren-mosaic—see mosaic-*schlieren*
schlieren texture, nematic   35, 57
schlieren texture, S_C 57, 58, 59
short-range order   95, 97, 98
smectics, structural features, diagram of   153
smectics, structural features, tabulation   154, 155, 156
smectic
  —A   1
  —A_1 143, 144
  —A_2 145
  —A_d 143
  —Ã   145, 146
  —AB transition   20
  —B   23
  —B (crystal and hexatic)   24, 25, 134
  —B_2 149
  —B_A 24

Smectic—*continued*
  —B_C 24
  —C   45
  —C_1 146
  —C_2 146
  —C̃   146, 147
  —D   68
  —E   82
  —F   94, 151, 152
  —G   23, 105
  —H   23, 105, 120
  —I   127
  —J   110, 140, 151, 153
  —K   121, 142, 150, 153
spherulitic texture   148
spontaneous polarization   62
stacking arrangements   31
stacking faults   31
stacking rearrangements   31
stepped droplets   1
stepped edges   39
streaks, smectic A   19
structural features of smectics
  diagram   153
  tables of   154, 155, 156
surfactants   8
symmetry, molecular   56

terminal outboard dipole   49
three-dimensional bond-orientational order   96
three-dimensional hexatic   135
three-dimensional order   29
tilt angle   45, 47
tilt angle variation   47
tilt correlation   46
tilt direction   45, 96, 97, 108, 109, 121
tilt distribution   4, 5
tilted antiphase   146, 147
tilted hexatic phases   139
tilted smectic B   105
tilted smectic E   106, 120, 121
topology, point singularity   58, 59
transition bars   39, 139
transverse dipoles   49, 53, 56
transverse modulation   29, 33, 34, 146
trilayer stacking   31
truncated fan   39
two-dimensional structure, smectic F   96, 97
two-dimensional crystal phase   130
two-dimensional hexatic phase   134

uniaxial phase, S_E 82, 83
unified nomenclature system   106
uncorrelated layers   130

volume change   140

wave motion 33, 34
walls in smectic mosaics 37, 38
wishbones, parabolic focal-conic
  defects 39, 139
Wulf's model for smectic C 51

X-ray diffraction
  —A 21
  —B 25, 43, 136
  —C 66
  —D 80

X-ray diffraction—*continued*
  —E 83, 93
  —F 95, 104
  —G 33, 118, 121
  —H 118, 126
  —I 129, 133
  —J 121
  —K 121

zig-zag shaped molecules 52

Plates 1-124

Plate 1 The focal-conic fan texture of the smectic A phase of n-decyl 4-(4'-phenylbenzylideneamino) cinnamate.

Plate 2 The separation of the smectic A phase in the form of bâtonnets from the isotropic liquid of diethyl 4,4'-azoxydibenzoate.

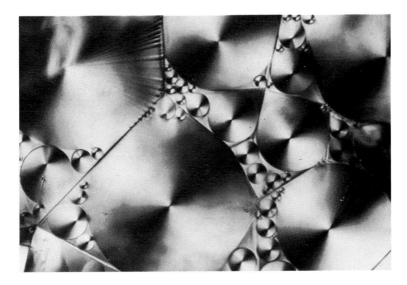

Plate 3 The polygonal texture of the smectic A phase of n-butyl 4-(4'-phenylbenzylideneamino) cinnamate.

Plate 4 The natural texture of the smectic B phase obtained on cooling the isotropic liquid of 4-n-hexyl-4'-n-hexyloxybiphenyl; the black areas are homeotropic S_B.

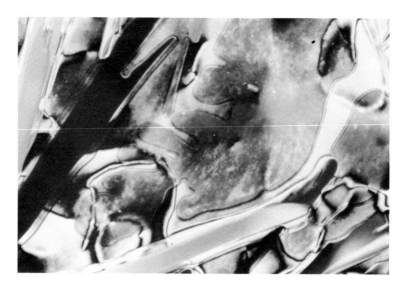

Plate 5 The transition to the mosaic texture of the smectic B phase on cooling the nematic phase of 4-n-pentyloxybenzylidene-4'-aminobiphenyl.

Plate 6 The mosaic texture of the smectic B phase of 4-n-pentyloxybenzylidene-4'aminobiphenyl (oblong platelets).

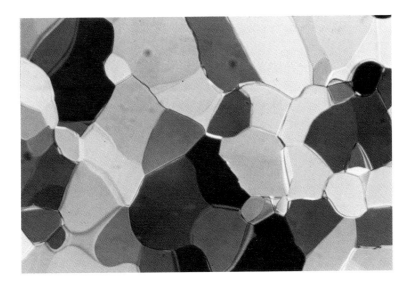

Plate 7 The mosaic texture of the smectic B phase of n-propyl 4-(4′-propylmercaptobenzylidene-amino)cinnamate (more rounded platelets).

Plate 8 The paramorphotic focal-conic fan texture of the ordered smectic B phase of n-decyl 4-(4′-phenylbenzylideneamino)cinnamate formed on cooling the smectic A phase shown in Plate 1.

Plate 9 The paramorphotic focal-conic fan texture of the hexatic, fluid smectic B phase of n-hexyl 4′-n-pentyloxybiphenyl-4-carboxylate.

Plate 10 The paramorphotic truncated focal-conic fan texture of the smectic B phase formed on cooling the smectic A phase of N-(4′-n-heptyloxybenzylidéne)-4-ethylaniline (70.2).

Plate 11 The transition from the focal-conic fan texture of the smectic A phase to the paramorphotic focal-conic fan texture of the smectic B phase for n-decyl 4-(4′-phenylbenzylideneamino) cinnamate. The backs of the fans are crossed with transition bars.

Plate 12 More pronounced transition bars occurring at the smectic A to smectic B phase change in a mixture of terephthalylidene-bis-4-n-butylaniline (50% by wt) and n-hexyl 4′-n-dodecyloxybiphenyl-4-carboxylate (50% by wt).

Plate 13 The *paramorphotic* focal-conic fan texture (also showing some areas with a moss-like appearance) of the smectic B phase formed after the isotropic liquid to S_{AB} transition for 3-methylbenzyl 4-(4'-phenylbenzylideneamino)cinnamate.

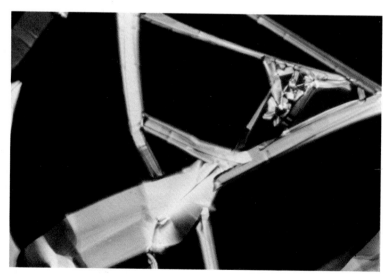

Plate 14 The *separation* of the smectic B phase from the isotropic liquid *via* an infinitely short smectic A phase for 3-methylbenzyl 4-(4'-phenylbenzylideneamino)cinnamate — compare with Plate 13 which shows the texture on completion of the transition.

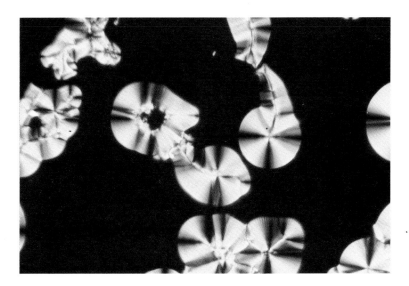

Plate 15 The transition from the isotropic liquid to the S_A to the S_B phase for methyl 4'-n-octyloxybiphenyl-4-carboxylate. The S_A phase has a small, but more finite temperature range than that for the ester in Plate 13 and 14.

Plate 16 The natural texture of the smectic B phase separating from the nematic phase of *trans,trans*-4-n-propylbicyclohexyl-4'-carbonitrile on slow cooling.

Plate 17 The large platelets of the mosaic texture of the smectic B phase formed on cooling the nematic phase of *trans,trans*-4-n-propylbicyclohexyl-4′-carbonitrile.

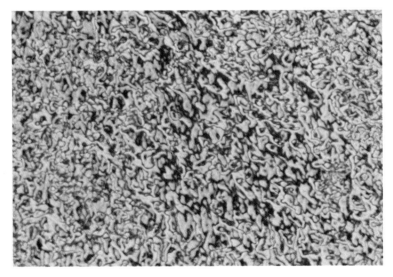

Plate 18 The natural *schlieren* texture of the smectic C phase formed on cooling the nematic phase of ($\pm$)-4-(2′-methylbutyl)phenyl 4′-(4″-methylhexyl)biphenyl-4-carboxylate.

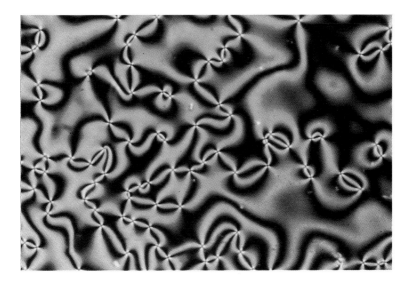

Plate 19 The *schlieren* texture of the smectic C phase formed on cooling the homeotropic texture of the smectic A phase of 4-n-hexyloxyphenyl 4′-n-octyloxybiphenyl-4-carboxylate. The texture exhibits only centres with four *schlieren*.

Plate 20 The sanded texture of the smectic C phase of 3′-nitro-4′-n-hexadecyloxybiphenyl-4-carboxylic acid.

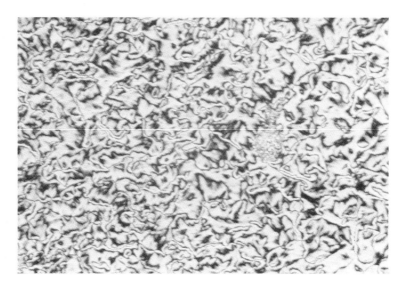

Plate 21 The *schlieren* texture of the smectic C phase of 4-n-hexyloxyphenyl 4′-n-octyloxybiphenyl-4-carboxylate. The small domains exhibit *faint* lines caused by layer stress as the tilt angle changes.

Plate 22 The paramorphotic broken focal-conic fan texture of the smectic C phase formed on cooling the smectic A phase of terephthalylidene-bis-4-n-butylaniline.

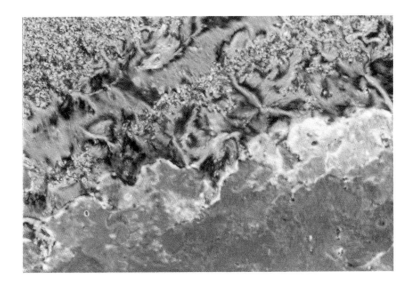

Plate 23 Contact preparation of chiral and non-chiral smectic C phases. A racemic mixture of two optically active isomers of 4-(2'-methylbutyl)phenyl 4'-n-octyloxybiphenyl-4-carboxylate (top of plate) is allowed to form a contact preparation with S-(+)-4-(2'-methylbutyl)phenyl 4'-n-octyloxybiphenyl-4-carboxylate (lower part of plate).

Plate 24 The petal texture of the chiral smectic C phase of S-(+)-4-n-hexyloxyphenyl 4'-(4''-methylhexyl)biphenyl-4-carboxylate.

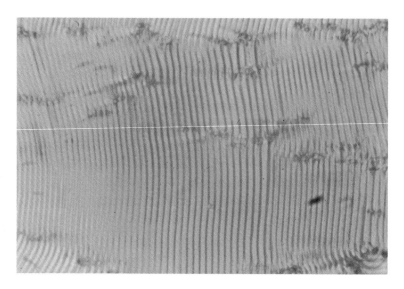

Plate 25 The texture of the chiral smectic C phase when the plane layers dip at an angle to the glass surface for S-(+)-terephthalylidene-bis-4-(4'-methylhexyloxy)aniline. The equally spaced lines are related to the pitch of the phase.

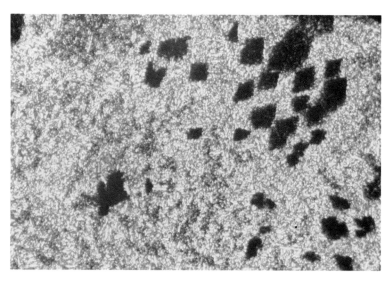

Plate 26 Optically isotropic areas of the smectic D phase forming in the sanded texture of the smectic C phase on heating 3'-nitro-4'-n-hexadecyloxybiphenyl-4-carboxylic acid.

Plate 27 The fern growth pattern of the smectic D phase forming in the *schlieren* C texture of 3'-nitro-4'-n-hexadecyloxybiphenyl-4-carboxylic acid.

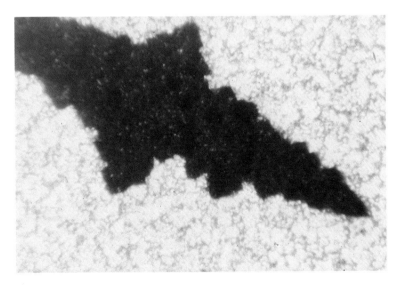

Plate 28 The fern-like growth pattern of the smectic D phase of 3'-nitro-4'-n-hexadecyloxybiphenyl-4-carboxylic acid. The fern pattern demonstrates the growth of the cubic or octahedral domains as they mesh together to form a single domain.

Plate 29 The texture of the S_4 or discotic phase formed on cooling the smectic A phase of 3'-nitro-4'-n-hexadecyloxybiphenyl-4-carboxylic acid.

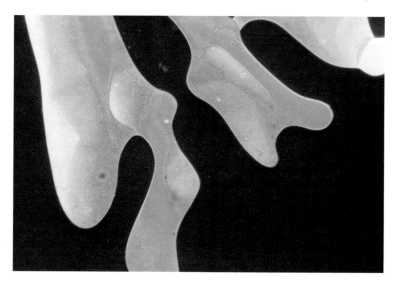

Plate 30 The natural texture of the smectic E phase separating from the isotropic liquid on cooling 4-ethoxy-4'-acetylbiphenyl.

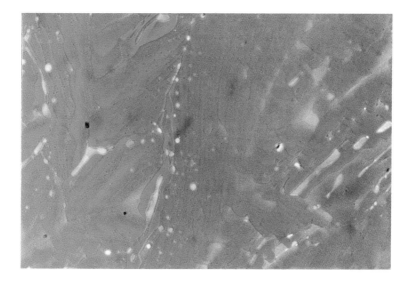

Plate 31 The natural mosaic texture of the smectic E phase obtained on cooling the isotropic liquid of 4-ethoxy-4'-acetylbiphenyl.

Plate 32 The paramorphotic arced focal-conic fan texture of the smectic E phase obtained on cooling the focal-conic textures of the smectic A and smectic B phases of methyl 4'-n-octyloxybiphenyl-4-carboxylate (see also Sequence 7).

Plate 33 The platelet texture of the smectic E phase obtained on cooling the homeotropic texture of the smectic A phase of di-n-propyl *p*-terphenyl-4,4″-carboxylate.

Plate 34 The paramorphotic arced focal-conic fan texture of the smectic E phase (again showing moss-like areas) obtained on cooling the focal-conic-mosaic texture obtained after the isotropic liquid to S_{AB} transition for 3-methylbenzyl 4-(4′phenylbenzylideneamino) cinnamate.

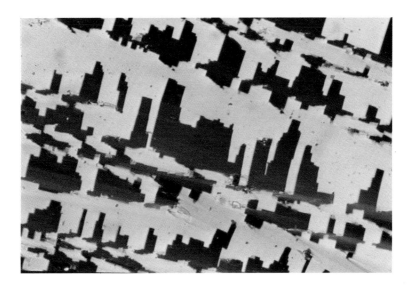

Plate 35 The paramorphotic broken focal-conic fan texture of the smectic F phase formed on cooling the focal-conic texture of the smectic C phase of terephthalylidene-bis-4-n-pentylaniline (TBPA).

Plate 36 The paramorphotic *schlieren*-mosaic texture of the smectic F phase formed on cooling the *schlieren* texture of the smectic C phase of terephthalylidene-bis-4-n-pentylaniline (TBPA).

Plate 37 The paramorphotic broken focal-conic fan texture of the smectic F phase formed on cooling the smectic A phase of N-(4-n-nonyloxybenzylidene)-4′-n-butylaniline (90.4).

Plate 38 The natural mosaic texture of the smectic F phase formed on cooling the homeotropic texture of the smectic A phase of N-(4-n-nonyloxybenzylidene)-4′-n-butylaniline (90.4)

Plate 39 The very fine mosaic texture of the smectic F phase of N-(4-n-pentyloxybenzylidene)-4′-n-hexylaniline (50.6) formed on cooling the homeotropic texture of the preceding smectic B phase.

Plate 40 The broken focal-conic texture of the chiral smectic F phase of (+)-4-(2″-chlorobutanoyloxy)-4′-n-octyloxybiphenyl.

Plate 41 The *schlieren*-mosaic texture of the chiral smectic F phase of (+)-4-(2″-chlorobutanoyloxy)-4′-n-octyloxybiphenyl.

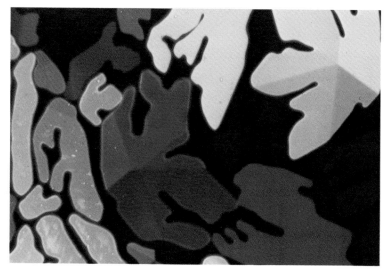

Plate 42 The smectic G phase of (±)-4-(2″-chlorobutanoyloxy)-4′-n-pentyloxybiphenyl separating from the isotropic liquid on cooling.

Plate 43 The natural mosaic texture of the smectic G phase of (±)-4-(2″-chlorobutanoyloxy)-4′-n-pentyloxybiphenyl formed on cooling the isotropic liquid.

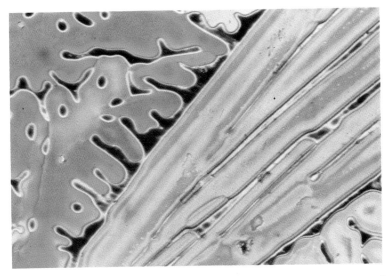

Plate 44 The mosaic texture of the smectic G phase separating from the nematic phase of N-(4-n-butyloxybenxylidene)-4′-ethylaniline (40.2) on rapid cooling.

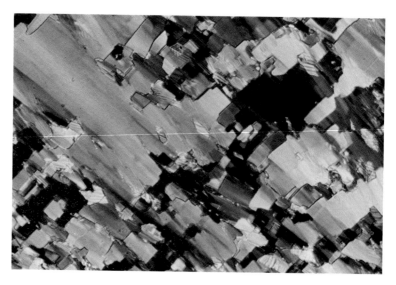

Plate 45 The paramorphotic broken focal-conic fan texture of the smectic G phase formed on cooling the fan texture of the smectic C phase of terephthalylidene-bis-4-n-butylaniline (TBBA).

Plate 46 The mosaic texture of the smectic G phase formed on cooling the *schlieren* texture of the smectic C phase of terephthalylidene-bis-4-n-butylaniline (TBBA).

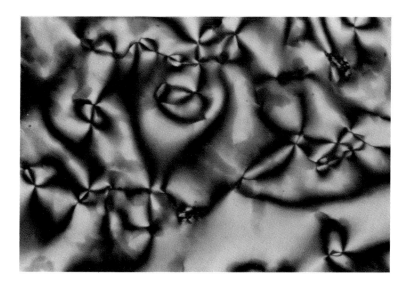

Plate 47 The texture of the 'chiral' smectic G phase formed on cooling the *schlieren* texture of the chiral smectic C phase of (+)-4-(2'-methylbutyl)phenyl 4'-n-octyloxybiphenyl-4-carboxylate. This texture is of a mosaic kind, but retains *schlieren* characteristics from the smectic C phase.

Plate 48 Immiscibility of the smectic G mosaic texture and the smectic B homeotropic texture in a thoroughly mixed, two-component mixture of terephthalylidene-bis-4-n-butylaniline (80% by wt) (S_G) and 4-n-pentyloxyphenyl 4'-n-octyloxybipenyl-4-carboxylate (20% by wt) (S_B). The two phases segregate on cooling the continuous and homogeneous *schlieren* texture of the preceding smectic C phase.

Plate 49 The paramorphotic broken focal-conic fan and mosaic textures of the smectic H'(K) phase formed on cooling the fan and mosaic textures of the smectic G phase of (±)-4-(2'-methylbutyl)phenyl 4'-n-decyloxybiphenyl-4-carboxylate.

Plate 50 The paramorphotic broken focal-conic fan texture of the smectic H'(K) phase of (±)-4-(2'-methylbutyl)phenyl 4'-n-octyloxybiphenyl-4-carboxylate.

Plate 51 The mosaic texture of the smectic H phase with zig-zag lines formed from the mosaic texture of the smectic G phase of terephthalylidene-bis-4-n-butylaniline (TBBA).

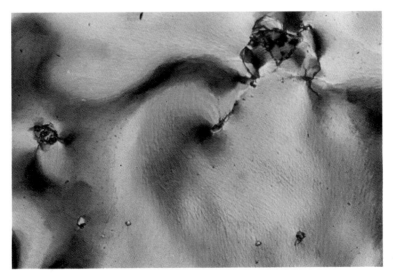

Plate 52 The grained paramorphotic mosaic texture of the smectic H phase formed on cooling the mosaic texture of the smectic G phase of terephthalylidene-bis-4-n-pentylaniline (TBPA).

Plate 53 The natural texture of the smectic I phase formed on cooling the isotropic liquid of 4,4'-bis-(n-octadecylamino)biphenyl.

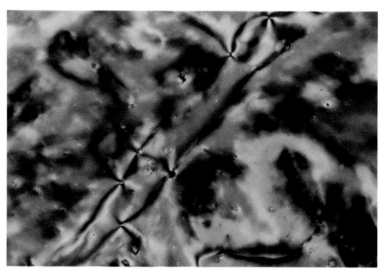

Plate 54 The *schlieren* texture of the smectic I phase formed on cooling the *schlieren* texture of the smectic C phase of (±)-4-(2'-methylbutyl)phenyl 4'-n-nonyloxybiphenyl-4-carboxylate.

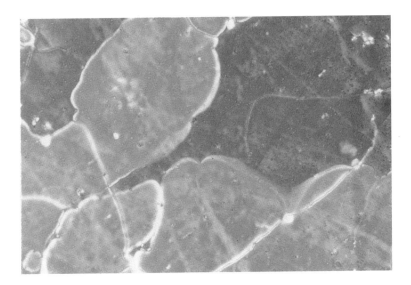

Plate 55 The bubble or plane texture of the chiral smectic I phase of (+)-4-(2′-methylbutyl)phenyl 4′-n-octyloxybiphenyl-4-carboxylate formed on cooling the plane texture the preceding chiral smectic C phase.

Plate 56 The focal-conic texture of the chiral smectic I phase of (+)-4-(2′-methylbutyl)phenyl 4′-n-octyloxybiphenyl-4-carboxylate.

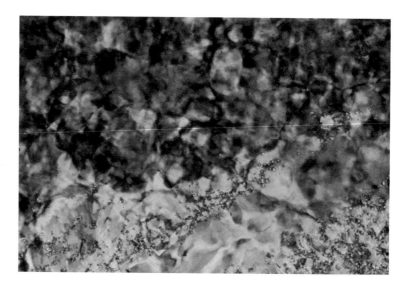

Plate 57 A contact preparation between the optically active smectic I phase and its achiral counterpart. The material is racemic and S-(+)-4-(2′-methylbutyl)phenyl 4′-n-octyloxybiphenyl-4-carboxylate.

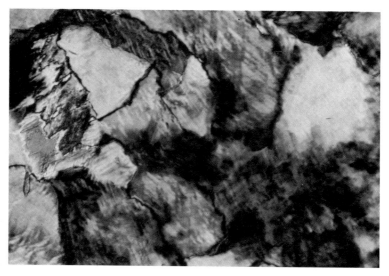

Plate 58 Unidentified S_{VII} phase of terephthalylidene-bis-4-n-butylaniline (TBBA) formed on cooling the mosaic texture of the preceding smectic H phase.

Plate 59 The unidentified S_2 phase of *trans,trans*-4-n-propylbicyclohexyl-4′-carbonitrile formed on cooling the mosaic texture of the preceding S_B phase.

Plate 60 The second unidentified S_3 phase of *trans,trans*-4-n-propylbicyclohexyl-4′-carbonitrile formed on cooling the mosaic texture of the preceding S_2 phase.

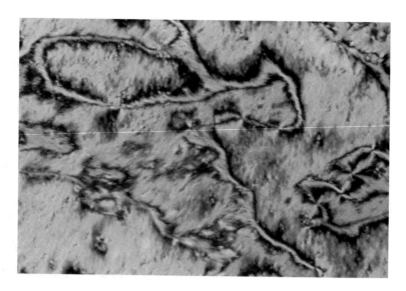

Plate 61 *Sequence 1.* The *schlieren* texture of the smectic C phase of racemic 4-(2′-methylbutyl)phenyl 4′-n-octyloxybiphenyl-4-carboxylate.

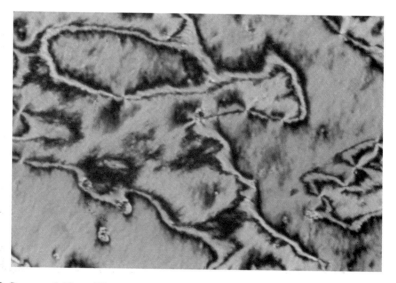

Plate 62 *Sequence 1.* The *schlieren* texture of the smectic C phase on further cooling — note the colour changes as the tilt angle increases with decreasing temperature — for racemic 4-(2′-methylbutyl)phenyl 4′-n-octyloxybiphenyl-4-carboxylate.

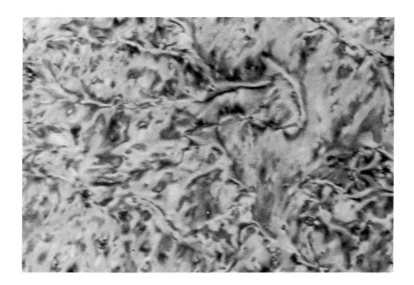

Plate 63 *Sequence 1.* The S_C to S_I phase transition on cooling for racemic 4-(2′-methylbutyl)phenyl 4′-n-octyloxybiphenyl-4-carboxylate (same area as Plates 61 and 62).

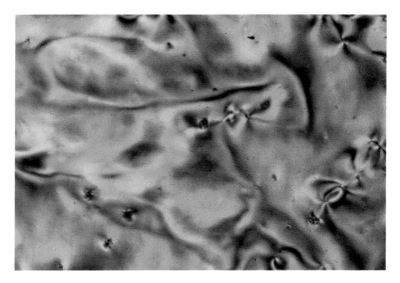

Plate 64 *Sequence 1.* The *schlieren* texture of the smectic I phase of racemic 4-(2′-methylbutyl)phenyl 4′-n-octyloxybiphenyl-4-carboxylate (same area as previous Plates).

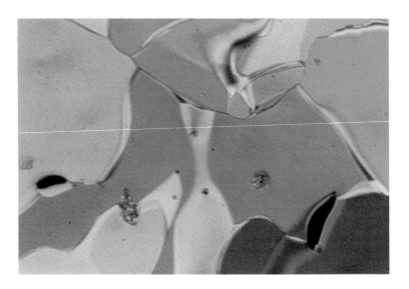

Plate 65 *Sequence 1.* The mosaic texture of the smectic G'(J) phase of racemic 4-(2'-methylbutyl)phenyl 4'-n-octyloxybiphenyl-4-carboxylate (same area as previous Plates).

Plate 66 *Sequence 1.* The transition from the mosaic texture of the smectic G'(J) phase to the smectic H'(K) phase of racemic 4-(2'-methylbutyl)phenyl 4'-n-octyloxybiphenyl-4-carboxylate (same area as Plate 65).

Plate 67 *Sequence 1.* The mosaic texture of the smectic H'(K) phase of racemic 4-(2'-methylbutyl)phenyl 4'-n-octyloxybiphenyl-4-carboxylate (same area as Plate 66).

Plate 68 *Sequence 2.* The focal-conic fan texture of the smectic A phase of S-(+)-4-(2'-methylbutyl)phenyl 4'-n-octyloxybiphenyl-4-carboxylate.

Plate 69 *Sequence 2.* The banded focal-conic fan texture of the chiral smectic C phase of S-(+)-4-(2'-methylbutyl)phenyl 4'-n-octyloxybiphenyl-4-carboxylate.

Plate 70 *Sequence 2.* The broken focal-conic fan texture of the chiral smectic I phase of S-(+)-4-(2'-methylbutyl)phenyl 4'-n-octyloxybiphenyl-4-carboxylate.

Plate 71 *Sequence 2*. The broken, banded focal-conic fan texture of the 'chiral' smectic G'(J) phase of S-(+)-4-(2'-methylbutyl)phenyl 4'-n-octyloxybiphenyl-4-carboxylate (same area as Plates 68, 69, and 70).

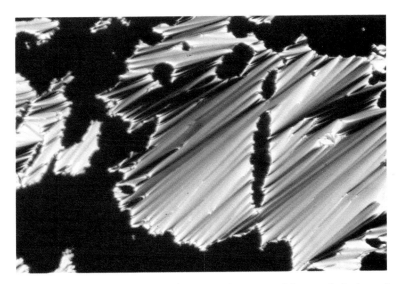

Plate 72 *Sequence 3*. The focal-conic fan and homeotropic textures of the smectic A phase of N-(4-n-heptyloxybenzylidene)-4'-n-pentylaniline (70.5).

Plate 73 *Sequence 3.* The broken focal-conic fan and *schlieren* textures of the smectic C phase of *N*-(4-n-heptyloxybenzylidene)-4'-n-pentylaniline (70.5).

Plate 74 *Sequence 3.* The focal-conic fan and homeotropic textures of the crystal smectic B phase of *N*-(4-n-heptyloxybenzylidene)-4'-n-pentylaniline (70.5).

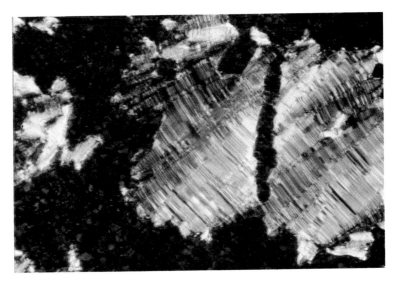

Plate 75 *Sequence 3.* The arced or banded focal-conic fan and mosaic textures of the smectic G phase of *N*-(4-n-heptyloxybenzylidene)-4′-n-pentylaniline (70.5).

Plate 76 *Sequence 4.* The focal-conic fan texture of the crystal smectic B phase of *N*-(4-n-pentyloxybenzylidene)-4′-n-hexylaniline (50.6).

Plate 77 *Sequence 4.* The broken focal-conic fan texture of the smectic F phase of *N*-(4-n-pentyloxybenzylidene)-4′-n-hexylaniline (50.6).

Plate 78 *Sequence 4.* The broken focal-conic fan texture of the smectic G phase of *N*-(4-n-pentyloxybenzylidene)-4′-n-hexylaniline (50.6).

Plate 79 *Sequence 5.* The *schlieren* and focal-conic fan textures of the smectic I phase formed on cooling the C phase of N,N'-bis-(4′-n-heptyloxybenzylidene)-1,4-phenylenediamine. Note the focal-conic fan texture is relatively unbroken because the C phase was formed directly from the nematic phase. This textural property is carried through all of the phases (Plates 80, 81, and 82).

Plate 80 *Sequence 5.* The mosaic and focal-conic fan textures of the smectic G′(J) phase formed on cooling the smectic I phase of N,N'-bis-(4′-n-heptyloxybenzylidene)-1,4-phenylenediamine.

Plate 81 *Sequence 5.* The transition textures at the smectic G'(J) to smectic H'(K) phase change of N,N'-bis-(4'-n-heptyloxybenzylidene)-1,4-phenylenediamine.

Plate 82 *Sequence 5.* The mosaic and focal-conic fan textures of the smectic H'(K) phase of N,N'-bis-(4'-n-heptyloxybenzylidene)-1,4-phenylenediamine.

Plate 83 *Sequence 6.* The focal-conic fan and homeotropic textures of the smectic A phase of terephthalylidene-bis-4-n-decylaniline (TBDA).

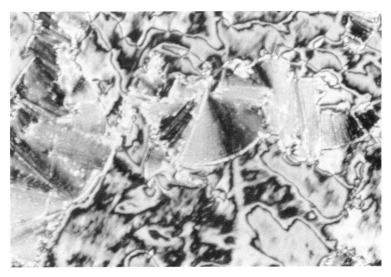

Plate 84 *Sequence 6.* The *schlieren* and broken focal-conic fan textures of the smectic C phase of terephthalylidene-bis-4-n-decylaniline (TBDA).

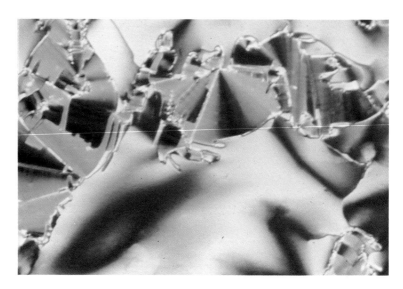

Plate 85 *Sequence 6.* The *schlieren* and focal-conic fan textures of the smectic I phase of terephthalylidene-bis-4-n-decylaniline (TBDA).

Plate 86 *Sequence 6.* The mosaic and focal-conic fan textures of the smectic F phase of terephthalylidene-bis-4-n-decylaniline (TBDA).

Plate 87 *Sequence 7.* The smectic A phase of n-hexyl 4'-n-pentyloxybiphenyl-4-carboxylate (65OBC), formed on cooling the isotropic liquid.

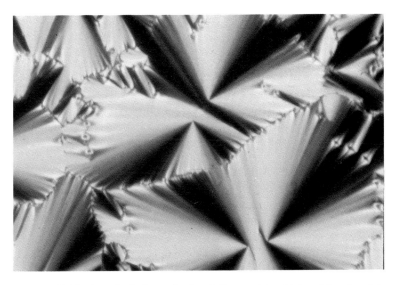

Plate 88 *Sequence 7.* The hexatic B phase of n-hexyl 4'-n-pentyloxybiphenyl-4-carboxylate (65OBC) formed on cooling the focal-conic texture of the preceding A phase.

Plate 89 *Sequence 7.* The smectic E phase of n-hexyl 4'-n-pentyloxybiphenyl-4- carboxylate (65OBC) typically showing a banded focal-conic texture.

Plate 90 *Sequence 7.* The smectic A phase of n-hexyl 4'-n-pentyloxybiphenyl-4-carboxylate (65OBC) formed on heating the hexatic B phase. The backs of the focal-conic domains become crossed with parabolic focal-conic defects (wishbones) — see also Sequence 8.

Plate 91 *Sequence 8.* The focal-conic fan texture of the smectic A phase of n-hexyl 4′-n-pentyl-biphenyl-4-thiolcarboxylate (65SBC).

Plate 92 *Sequence 8.* The transition from the A phase to the crystal B phase of n-hexyl 4′-n-pentylbiphenyl-4-thiolcarboxylate (65SBC). The fan backs become crossed with transition bars.

Plate 93 *Sequence 8.* The focal-conic fan texture of the crystal B phase of n-hexyl 4'-n-pentyl-biphenyl-4-thiolcarboxylate (65SBC).

Plate 94 *Sequence 8.* The arced focal-conic fan texture of the smectic E phase formed on cooling the crystal B phase of n-hexyl 4'-n-pentylbiphenyl-4-thiolcarboxylate (65SBC).

Plate 95 *Sequence 8.* The smectic A phase of n-hexyl 4'-n-pentylbiphenyl-4-thiolcarboxylate (6SBC) formed on heating the crystal B phase. The backs of the focal-conic domains become patterned with parabolic focal-conic defects.

Plate 96 The crystal B phase of n-propyl 4'-n-pentylbiphenyl-4-thiolcarboxylate (3SBC) formed on cooling the focal-conic fan texture of the preceding A phase. Note the fans now have a stepped appearance; compare Plate 88 for the hexatic B phase.

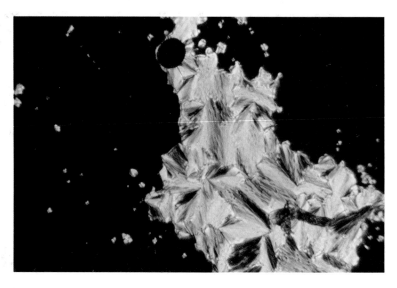

Plate 97 *Sequence 9.* The homeotropic and focal-conic fan textures of the bilayer smectic A phase of 4'-n-nonyloxy-4-biphenylyl 4-cyanobenzoate.

Plate 98 *Sequence 9.* The transition from the bilayer smectic A phase to the tilted antiphase ($S_{\tilde{C}}$) of 4'-n-nonyloxy-4-biphenylyl 4-cyanobenzoate.

Plate 99 *Sequence 9.* The textures of the tilted antiphase (S$\tilde{C}$) of 4'-n-nonyloxy-4-biphenylyl 4-cyanobenzoate.

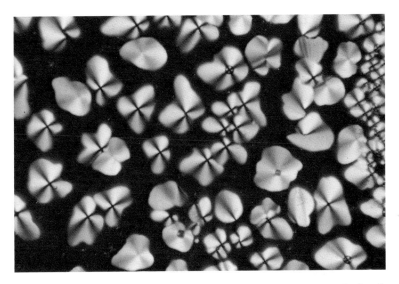

Plate 100 The transition from the homeotropic texture of the bilayer smectic A phase to the domain texture of the tilted antiphase (S$\tilde{C}$) of 4'-n-nonyloxy-4-biphenylyl 4-cyanobenzoate.

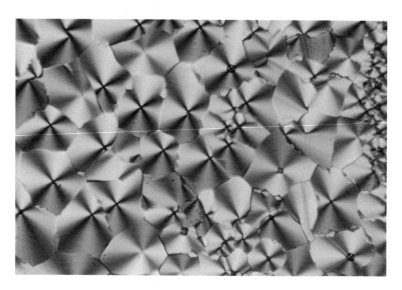

Plate 101 The domain texture of the tilted antiphase (S$\tilde{C}$) of 4'-n-decyloxy-4-biphenylyl 4-cyanobenzoate.

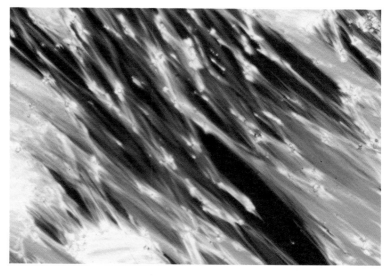

Plate 102 *Sequence 10.* The smectic A phase of 4-n-heptylphenyl 4-(4'-nitrobenzoyloxy)benzoate.

Plate 103 *Sequence 10.* The transition from the smectic A phase to the smectic A antiphase ($S_{\tilde{A}}$) of 4-n-heptylphenyl 4-(4'-nitrobenzoyloxy)benzoate.

Plate 104 *Sequence 10.* The smectic A antiphase ($S_{\tilde{A}}$) of 4-n-heptylphenyl 4-(4'-nitrobenzoyloxy)benzoate.

Plate 105 The smectic F phase separating from the isotropic liquid of 4-n-dodecanoyloxy-4′-n-octyloxybiphenyl.

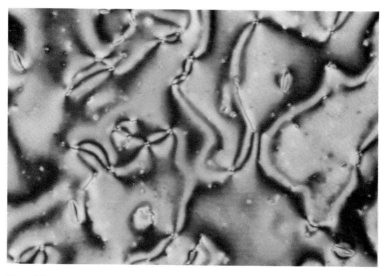

Plate 106 The *schlieren* texture of the nematic phase of 4-(2′-methylbutyl)phenyl 4′-(4″-methylhexyl)biphenyl-4-carboxylate.

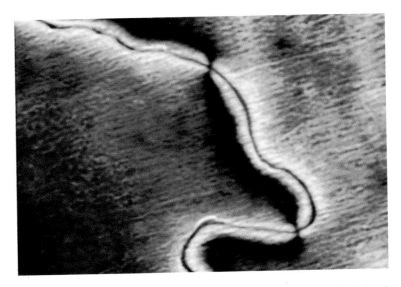

Plate 107 The threaded texture of the nematic phase of 4-(*trans*-4'-n-pentylcyclohexylmethoxy)benzonitrile showing two brushes originating from the centres.

Plate 108 The *schlieren* nematic texture of 4-(*trans*-4'-n-pentylcyclohexylmethoxy)benzonitrile showing four brushes radiating from several centres.

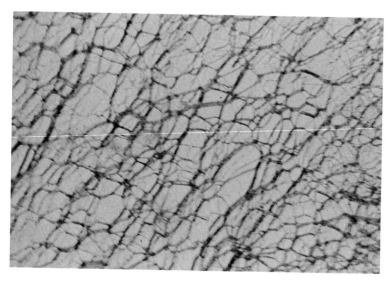

Plate 109 Plane texture of the cholesteric phase (with discontinuities) of chiral di-2-methylbutyl terephthalylidene-bis-4'-aminobenzoate.

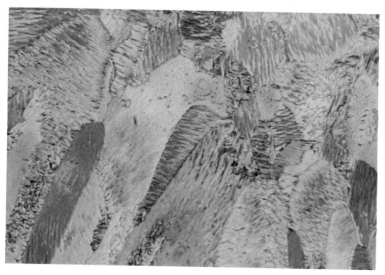

Plate 110 The focal-conic texture of the cholesteric phase of chiral 2-methylbutyl 4-(4'-nitrobenzylideneamino)cinnamate.

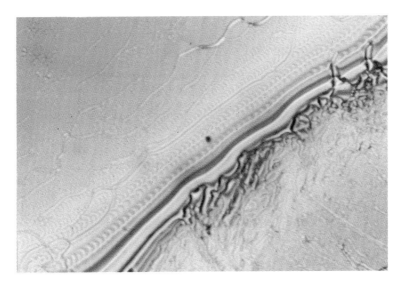

Plate 111 Rivulet of nematic in a contact preparation of the plane textures of two cholesteric materials of opposite pitch sense.

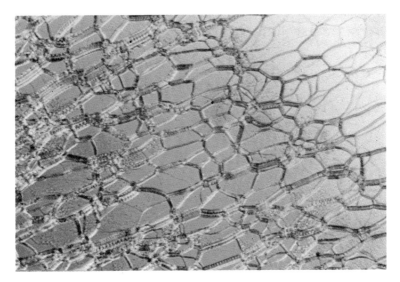

Plate 112 Continuity in a contact preparation of the plane textures of two cholesteric materials of the same pitch sense.

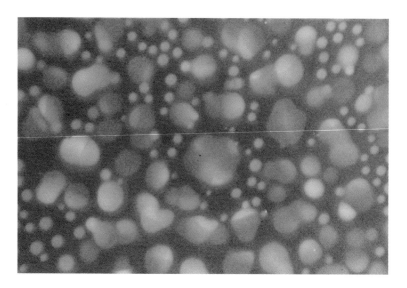

Plate 113 The blue fog phase of (+)-4'-n-hexyloxy-4-biphenylyl 4-(2'-methylbutyl)benzoate separating from the isotropic liquid.

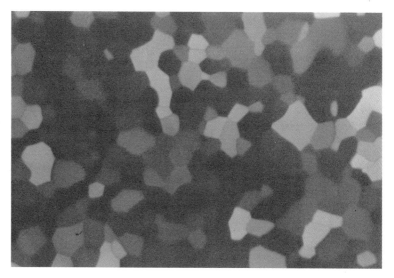

Plate 114 The platelet texture of 'blue' phase (II) of (+)-4'-n-hexyloxy-4-biphenylyl 4-(2'-methylbutyl)benzoate formed on cooling the blue fog phase.

Plate 115 The wrinkled platelet texture of 'blue' phase (I) of (+)-4'-n-hexyloxy-4-biphenylyl 4-(2'-methylbutyl)benzoate formed on cooling 'blue' phase (II).

Plate 116 The zig-zag blade texture of the cholesteric 'blue' phase of (+)-4-(2'-methylbutyl)phenyl 4'-(4"-methylhexyl)biphenyl-4-carboxylate.

Plate 117 The discotic phase of benzene-hexa-n-heptanoate separating from the isotropic liquid on cooling.

Plate 118 The discotic phase of benzene-hexa-n-heptanoate reverting to the isotropic liquid on heating.

Plate 119 The pseudo focal-conic texture of the discotic phase of benzene-hexa-n-heptanoate.

Plate 120 The feather texture of the discotic phase of benzene-hexa-n-heptanoate.

Plate 121 The pseudo focal-conic fan texture of the discotic phase of di-isobutylsilanediol.

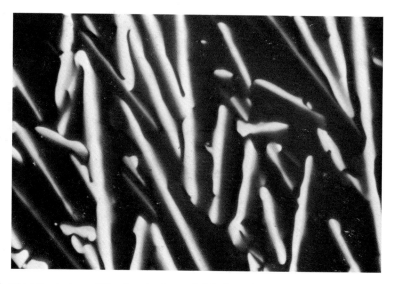

Plate 122 The spine texture of the discotic phase of di-isobutylsilanediol.

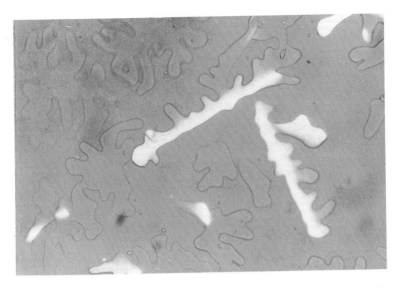

Plate 123 The spine texture of the discotic phase of di-isobutylsilanediol separating from the isotropic liquid.

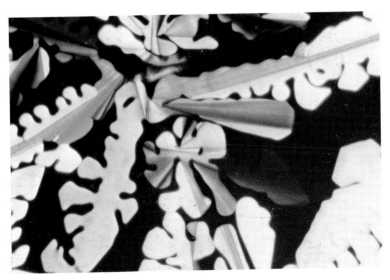

Plate 124 The texture of the discotic phase of Uroporphyrin (I) octa-n-dodecyl ester separating from the isotropic liquid.

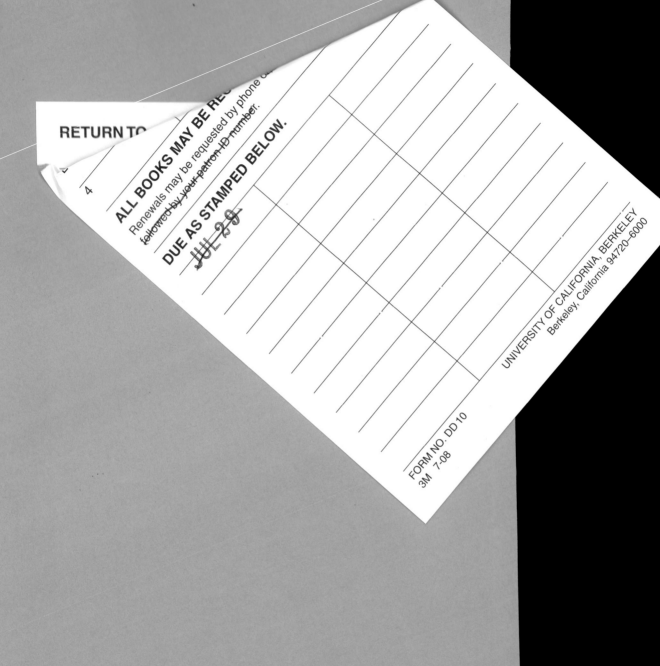